PRAISE FOR JOHN BRADSHAW'S *CAT SENSE*

"*Cat Sense* is jam-packed with fascinating (and contrarian) tidbits . . . covering everything from why cats purr to why they bring us dead things—and why we keep them around, even though their original purpose as mousers is mostly obsolete. Obligatory cat pun coming up . . . this book is a purrfect gift for the cat lady or cat dude in your life."—NPR, Book of the Year

"Bradshaw's book mixes pellets of cat lore with accounts of feline evolution, anatomy, genetics and development from newborn kitten to adulthood, plus descriptions of cat-psychology experiments in the laboratory, many of which he has conducted himself. . . . Inveterate cat-haters, those defective humans, probably won't appreciate this book, but anyone else might. It is written in a friendly and engaging way, has helpful tips for cat owners, and is packed with excellent cat facts."—*Guardian* (UK)

"[Bradshaw] starts with cat origins and works methodically—and illuminatingly—through the many daft anthropomorphic assumptions. . . . What makes Bradshaw's book so valuable is his positive thinking. How can we make the cat less anxious? How can we help? . . . [*Cat Sense* is] a mind-altering book."

—*Sunday Times* (UK)

"Bradshaw, who has been studying the behavior of domesticated animals for over 30 years, reveals some fascinating explanations for why cats act the way they do around humans."—*Time*

"Bradshaw . . . flags his seriousness of purpose with his subtitle, *How the New Feline Science Can Make You a Better Friend to Your Pet.* Bradshaw means to get into the cat brain. He's already plumbed its canine counterpart, in the 2011 book *Dog Sense*, which was also grounded in research, not sentiment, and in the idea that pets have inner lives more complicated than we imagine."

—Frank Bruni, *New York Times*

"Bradshaw is . . . a dedicated scientist, with much to teach us about our furry darlings."—*Telegraph* (UK)

"Books about animals tend to swing from how-to manuals devoid of evidence for the tactics they propose to scientific tracts with little comment on the way we actually live with our four-legged friends. *Cat Sense* strikes a nice balance, perhaps because Bradshaw researched it for 30 years. He synthesizes academic articles, experiments and his own observations into a lively, readable text."

—*Smithsonian*

"[Bradshaw] deftly sums up the latest science that attempts to discover what's going on inside the kitty brain. . . . A careful read can help a cat owner understand why cats don't get along, guide efforts in training and even reveal what's behind kitty's favorite toy."—*Science News*

"On physiology, Bradshaw goes well beyond charming did-you-knows to provide insights that could transform the average cat owner's understanding of their pet. . . . After reading *Cat Sense*, you will never look at your cat in the same way again."—*New Statesman* (UK)

"You could buy a dozen books by the many cat whisperers, cat gurus and cat therapists that exist in our feline-obsessed modern world, but their accumulated wisdom would probably not help you understand your cats . . . as well as *Cat Sense*."—*Observer* (UK)

"This is a fascinating book that reveals much new information. . . . Whether you share your home with a cat or just admire them from afar, this book is must reading. It is meticulously researched, crisply written, and an essential guide that offers penetrating insights about the domestic cat, many beliefs that will challenge our most basic assumptions but promise to dramatically improve not only the lives of our pets but ours as well."—*Tuscon Citizen*

"In his wide-ranging new book, *Cat Sense*, English anthrozoologist John Bradshaw calls on all his scientific resources to interpret our enigmatic felines for the 21st century—a restrictive era far removed from the predatory instincts of these not-quite-domesticated animals."—*Globe and Mail* (Toronto)

"[Bradshaw] offers plenty of insights into what makes your tabby purr and how those insights can make a difference in your domestic life. . . . The understanding you gain should make for a happier cat-human household."—*Natural History*

"If John Bradshaw's new book doesn't entirely penetrate the feline mystery . . . it does at least shine a beam of light on the question of what is going on in those furry little heads. Not all that much is often the answer. *Cat Sense* is an amiable and interesting round-up of the history and science of the domestic moggy, from its first appearance in prehistory to the latest behavioural and genetic discoveries."—*Literary Review* (UK)

"Bradshaw does a great job of explaining to the clueless cat owner what science has discovered about their pet. . . . [A] fascinating bookshelf essential for anyone who's ever looked at their cat and wondered what's going on behind those big eyes."—*Express* (UK)

"[I]nsightful. . . . Using cutting-edge research, Bradshaw takes us into the mysterious mind of the domestic cat, explaining the cat's nature and needs, and, in doing, so deepens our understanding of our wild housemates and improves our relationships with them."—*Modern Cat*

"[A] go-to cat guide in one easy read. . . . For cat lovers, this book gives a vital look into the perspective of the cat. . . . The insight this book provides will not only help cat companions better understand their pet, it will allow them to create an ideal living situation for their cat. Keeping your cat happy and stress-free will ensure a comfortable home for everyone."—*Global Animal*

"Using research, his background in anthrozoology and his personal experiences with cats, Bradshaw has written a scientific book that remains easily accessible to any cat owner. He admits up front there is still plenty to learn about the domestic cat, but *Cat Sense* is a solid starting point and a must for present owners and potential owners alike. Readers will be more aware of their companions' behaviors and what those behaviors mean in terms of the human-cat relationship, thereby creating a richer, more fulfilling connection for each."—*Shelf Awareness for Readers*

"Bradshaw deftly weaves together history, science, cat lore and some interesting predictions on the future evolution of cats as pets and members of human households. . . . Well-written and as readable as many novels, with graphs, illustrations, photos and boxed vignettes throughout, *Cat Sense* is a serious look at the science of our feline companions that most cat lovers, owners and breeders will find both educational and enjoyable. It is far more than the typical 'cat book.'"—*Cat Talk*

"*Cat Sense* goes a long way toward educating humans about their feline companions so that we can continue to enjoy them–and, perhaps more importantly, make them happier to be around us."—*Writer's Voice*

"This fascinating book will be a bible for cat owners."—*Booklist*, Starred Review

"With more than 30 years of experience studying animal behavior, [Bradshaw] is able to convey valuable information to cat owners, regardless of their experience with the species, that will assist them in providing the stable physical environment that cats crave, as well as promoting the healthiest of relationships between cat and owner. . . . [E]nlightening."—*Library Journal*

"[Bradshaw] engagingly synthesizes recent academic research about cats. . . . Readable, practical, and original, this is likely to become the go-to book for understanding cat behavior."—*Publishers Weekly*

"A useful guide to help cat lovers better understand their elusive pets."—*Kirkus*

"This fascinating book is one of the finest ever written about cats. There was hardly a page where I did not learn something new, and John Bradshaw's many practical suggestions are truly excellent. Any cat lover is bound to discover in it much that is useful, interesting, and entertaining."—Jeffrey Moussaieff Masson, author of *When Elephants Weep* and *The Nine Emotional Lives of Cats*

"John Bradshaw knows cats. *Cat Sense* brims with many details sure to engage owners and animal lovers alike—and possibly even felines."—Alexandra Horowitz, author of *Inside of a Dog: What Dogs See, Smell, and Know* and *On Looking: Eleven Walks with Expert Eyes*

"*Cat Sense* makes sense of cats for us from an entirely fresh perspective, using a combination of history, science, logic, and a heartfelt compassion, certainly inspired by his own lovely kitty, Splodge. John Bradshaw has given us all the gift of being able to truly comprehend our pussycats and offer them the unselfish and satisfying life we have never quite known how to do until this book opened our eyes."—Tracie Hotchner, author of *The Cat Bible* and host/ producer of *Dog Talk (and Kitties, Too!)* on NPR and *Cat Crazy* and *Cat Chat & Dog Talk* on the Radio Pet Lady Network

"Based on rock solid science, *Cat Sense* is a beautifully written whirlwind tour of all things feline—from the influence of ancient Egyptian funeral rites on the behavior of modern cats to practical tips on how to keep an indoor cat happy. I am so glad I read this book . . . and so is my cat Tilly!"—Hal Herzog, author of *Some We Love, Some We Hate, Some We Eat: Why It's So Hard To Think Straight About Animals*

CAT SENSE

How the New Feline Science Can Make You a Better Friend to Your Pet

John Bradshaw

BASIC BOOKS
A MEMBER OF THE PERSEUS BOOKS GROUP
New York

Published by Basic Books,
A Member of the Perseus Books Group

Pen-and-wash drawings by Alan Peters, © 2013.

First paperback edition published in 2014 by Basic Books.

Books published by Basic Books are available at special discounts for bulk purchases in the United States by corporations, institutions, and other organizations. For more information, please address the Special Markets Department at the Perseus Books Group, 2300 Chestnut Street, Suite 200, Philadelphia, PA 19103, or call (800) 810-4145, ext. 5000, or email special.markets@perseusbooks.com.

Designed by Trish Wilkinson
Set in Goudy Old Style

Library of Congress Cataloging-in-Publication Data

Bradshaw, John, 1950–
Cat sense : how the new feline science can make you a better friend to your pet / John Bradshaw.
pages cm
Includes bibliographical references and index.
ISBN 978-0-465-03101-6 (hardcover) — ISBN 978-0-465-04095-7 (e-book)
1. Cats—Behavior. 2. Cats—Psychology. 3. Human-animal relationships. 4. Cat owners. I. Title.
SF446.5.B725 2013
636.8—dc23 2013020749

ISBN 978-0-465-06496-0 (paperback)

10 9 8 7 6 5 4 3 2 1

Contents

To Splodge
(1988–2004)
A Real Cat

Dogs look up to us: cats look down on us.

—WINSTON CHURCHILL

When a man loves cats, I am his friend and comrade, without further introduction.

—MARK TWAIN

Splodge

Preface

What is a cat? Cats have intrigued people ever since they first came to live among us. Irish legend has it that "a cat's eyes are windows enabling us to see into another world"—but what a mysterious world that is! Most pet owners would agree that dogs tend to be open and honest, revealing their intentions to anyone who will pay them attention. Cats, on the other hand, are elusive: we accept them on their terms, but they in turn never quite reveal what those terms might be. Winston Churchill, who referred to his cat Jock as his "special assistant," famously once observed of Russian politics, "It is a riddle, wrapped in a mystery, inside an enigma; but perhaps there is a key"; he might as well have been talking about cats.

Is there a key? I'm convinced that there is, and that it can be found in science. I've shared my home with quite a few cats—and have become aware that "ownership" is not the appropriate term for this relationship. I've witnessed the birth of several litters of kittens, and nursed my elderly cats through their heartbreaking final declines into senility and ill health. I've helped with the rescue and relocation of feral cats, animals that literally wanted to bite the hand that fed them. Still, I don't feel that, on its own, my personal involvement with cats has taught me very much about what they are really like. Instead, the work of scientists—field biologists, archaeologists, developmental biologists, animal psychologists, molecular biologists, and anthrozoologists such as myself—has provided me with the pieces that, once assembled, begin to reveal the cat's true nature. We are still missing some pieces, but the definitive picture is emerging. This is an opportune moment to take stock of what we know, what is still to be discovered, and, most important, how we can use our knowledge to improve cats' daily lives.

Getting an idea of what cats are thinking does not detract from the pleasures of "owning" them. One theory holds that we can enjoy our pets' company only through pretending that they are "little people"—that we keep animals merely to project our own thoughts and needs onto them, secure in the knowledge that they can't tell us how far off the mark we are. Taking this viewpoint to its logical conclusion, forcing us to concede that our cats neither understand nor care what we say to them, we might suddenly find that we no longer love them. I do not subscribe to this idea. The human mind is perfectly capable of simultaneously holding two apparently incompatible views about animals, without one canceling the other out. The idea that animals are in some ways like and in others quite unlike humans lies behind the humor of countless cartoons and greetings cards; these simply would not be funny if the two concepts negated each other. In fact, quite the opposite: the more I learn about cats, both through my own studies and through other research, the more I appreciate being able to share my life with them.

Cats have fascinated me since I was a child. We had no cats at home when I was growing up, nor did any of our neighbors. The only cats I knew lived on the farm down the lane, and they weren't pets, they were mousers. My brother and I would occasionally catch intriguing glimpses of one of them running from barn to outhouse, but they were busy animals and not over-friendly to people, especially small boys. Once, the farmer showed us a nest of kittens among the hay bales, but he made no special effort to tame them: they were simply his insurance against vermin. At that age, I thought that cats were just another farm animal, like the chickens that pecked around the yard or the cows that were driven back to the barn every evening for milking.

The first pet cat I ever got to know was the polar opposite of these farm cats, a neurotic Burmese by the name of Kelly. Kelly belonged to a friend of my mother's who had bouts of illness, and no neighbor to feed her cat while she was hospitalized. Kelly boarded with us; he could not be let out in case he tried to run back home, he yowled incessantly, he would eat only boiled cod, and he was evidently used to receiving the undivided attention of his besotted owner. While he was with us, he spent most of his time hiding behind the couch, but within a few seconds of the telephone ringing, he would emerge, make sure that my mother's attention was occupied by the person on the other end of the line, and then sink his long Burmese canines deep into her calf. Regular callers became accustomed to the idea that twenty seconds in, the conversation would be interrupted by a scream and then a muttered curse. Understandably, none of us became particularly fond of Kelly, and we were always relieved when it was time for him to head back home.

Not until I had pets of my own did I begin to appreciate the pleasures of living with a normal cat—that is to say, a cat that purrs when it is stroked and greets people by rubbing around their legs. These qualities were probably also appreciated by the first people to give houseroom to cats thousands of years ago; such displays of affection are also the hallmark of tamed individuals of the African wildcat, the domestic cat's indirect ancestor. The emphasis placed on these qualities has gradually increased over the centuries. While most of today's cat owners value them for their affection above all else, for most of their history, domestic cats have had to earn their keep as controllers of mice and rats.

As my experience with domestic cats grew, so did my appreciation of their utilitarian origins. Splodge, the fluffy black-and-white kitten we bought for our daughter as compensation for having to relocate, quickly grew into a large, shaggy, and rather bad-tempered hunter. Unlike many cats, he was fearless in the face of a rat, even an adult. He soon learned that depositing a rat carcass on our kitchen floor for us to find when we came downstairs for breakfast was not appreciated, and after that he kept his predatory activities private—without, I suspect, giving the rats themselves any respite.

However brave he was against a rat, Splodge usually kept away from other cats. Every now and then, we would hear the cat-flap clatter as he arrived home in a tearing hurry, and a quick glance out the window would usually reveal one of the older cats in the neighborhood, glaring in the general direction of our back door. He had a favorite hunting area in the park nearby, but kept himself inconspicuous when traveling there and back. His diffidence toward other cats, especially males, was not just typical of many cats; it also exemplified a weakness in social skills that is perhaps the greatest difference between cats and dogs. Most dogs find it easy to get along with other dogs; cats generally find other cats a challenge. Yet many of today's owners expect their cats to accept other cats without question—either when they themselves wish to get a second cat, or when they decide to move, depositing their unsuspecting cat into what another cat thinks is its territory.

For cats, a stable social environment is not enough; they rely on their owners to provide a stable physical environment as well. Cats are fundamentally territorial animals that put down powerful roots in their surroundings. For some, their owner's home is all the territory they need. Lucy, another of my cats, showed no interest in hunting, despite being Splodge's great-niece; she barely strayed more than a dozen yards from the house—except when she came into season and disappeared over the garden wall for hours on end. Libby, Lucy's daughter and born in my home, was as brave a hunter as Splodge had been, but preferred to call the tomcats to her rather than go to them. Even though they were all related and all lived in the same house most of their lives, Splodge, Lucy, and Libby all had distinctive personalities, and if I learned one thing from observing them it was that no cat is completely typical: cats have personalities, just as humans do. This observation inspired me to study how such differences come about.

The transformation of the cat from resident exterminator to companion cohabiter is both recent and rapid, and—especially from the cat's perspective—evidently incomplete. Today's owners demand a different set of qualities from their cats than would have been the norm even a century ago. In some ways, cats are struggling with their newfound popularity. Most owners would prefer that their cats did not kill cute little birds and mice, and those people who are more interested in wildlife than in pets are becoming increasingly vocal in their opposition to the cat's predatory urges. Indeed, cats now probably face more hostility than at any time in the past two centuries. Can cats possibly shake off their legacy as humankind's vermin exterminator of choice, and in just a few generations?

Cats themselves are oblivious to the controversy caused by their predatory natures, but all too aware of the difficulties they encounter in their dealings with other cats. Their independence, the quality that makes cats the ideal low-maintenance pet, probably stems from their solitary origins, but it has left them poorly equipped to cope with many owners' assumptions that they should be as adaptable as dogs. Can cats become more flexible in their social needs, so that they are unfazed by the proximity of other cats, without compromising their unique appeal?

One of my reasons for writing this book is to project what the typical cat might be like fifty years from now. I want people to continue to enjoy the company of a delightful animal, but I'm not sure that the cat, as a species, is heading in the right direction. The more I've studied cats, from the wildest feral to the most cosseted Siamese, the more I've become convinced that we can no longer afford to take cats for granted: a more considered approach to cat keeping and cat breeding is necessary if we are to ensure their future.

Acknowledgments

I began studying cat behavior more than thirty years ago, first at the Waltham Center for Pet Nutrition, later at the University of Southampton, and now at the University of Bristol's Anthrozoology Institute. Much of what I've learned has come from painstaking observation of cats themselves: my own, my neighbors', cats in adoption centers, the family of cats that used to share the Anthrozoology Institute's offices, and many ferals and farm cats.

Compared to the large number of canine scientists, rather few academics specialize in feline science, and even fewer make the domestic cat the focus of their attention. Those I've had the privilege of working with and who've helped me to form my ideas about how cats see the world include Christopher Thorne, David Macdonald, Ian Robinson, Sarah Brown, Sarah Benge (*née* Lowe), Deborah Smith, Stuart Church, John Allen, Ruud van den Bos, Charlotte Cameron-Beaumont, Peter Neville, Sarah Hall, Diane Sawyer, Suzanne Hall, Giles Horsfield, Fiona Smart, Rhiann Lovett, Rachel Casey, Kim Hawkins, Christine Bolster, Elizabeth Paul, Carri Westgarth, Jenna Kiddie, Anne Seawright, Jane Murray, and others too numerous to list.

I've also learned a great deal from discussions with colleagues both at home and abroad, including the late Professor Paul Leyhausen, Dennis Turner, Gillian Kerby, Eugenia Natoli, Juliet Clutton-Brock, Sandra McCune, James Serpell, Lee Zasloff, Margaret Roberts and her colleagues at Cats Protection, Diane Addie, Irene Rochlitz, Deborah

Goodwin, Celia Haddon, Sarah Heath, Graham Law, Claire Bessant, Irene Rochlitz, Patrick Pageat, Danielle Gunn-Moore, Paul Morris, Kurt Kotrschal, Elly Hiby, Sarah Ellis, Britta Osthaus, Carlos Driscoll, Alan Wilson, and the late and much-missed Penny Bernstein. My thanks also to the University of Bristol's School of Veterinary Medicine, especially Professors Christine Nicol and Mike Mendl, and Drs. David Main and Becky Whay, for nurturing the Anthrozoology Institute and its research.

My studies of cats have relied on the cooperation of many hundreds of volunteer cat owners (and their cats!), to whom I will always be grateful. Much of our research would have been impossible without the unstinting assistance of the UK's re-homing charities, including the RSPCA, the Blue Cross, and St. Francis Animal Welfare, and I am especially grateful to Cats Protection for two decades of practical and financial assistance.

Summarizing nearly thirty years of research on cat behavior into a form intended to be appreciated by the average cat owner has not been an easy task. I have had expert guidance from Lara Heimert and Tom Penn, my editors at Basic and Penguin respectively, and my indefatigable agent Patrick Walsh. Thank you all.

As in my previous books, I've turned to my dear friend Alan Peters to bring some of the animals to life in the illustrations, and just as before, he's done me more than proud.

Finally, I must thank my family for their forbearance for my enforced absences in what my granddaughter Beatrice calls "Pops' office."

Introduction

The domestic cat is the most popular pet in the world today. Across the globe, domestic cats outnumber "man's best friend," the dog, by as many as three to one.[1] As more of us have come to live in cities—environments for which dogs are not ideally suited—cats have, for many, become the lifestyle pet of choice. About one-third of US households have one or more cats, and they are found in more than a quarter of UK families. Even in Australia, where the domestic cat is routinely demonized as a heartless killer of innocent endangered marsupials, about a fifth of households own cats. All over the world, images of cats are used to advertise all kinds of consumer goods, from perfume to furniture to confectionery. The cartoon cat "Hello Kitty" has appeared on more than 50,000 different branded products in more than sixty countries, netting her creators billions of dollars in royalties. Even though a significant minority of people—perhaps as many as one person in five—don't like cats, the majority who do show no sign of relinquishing even a fraction of their affection for their favorite animal.

Cats somehow manage to be simultaneously affectionate and self-reliant. Compared to dogs, cats are low-maintenance pets. They do not need training. They groom themselves. They can be left alone all day without pining for their owners as many dogs do, but they will nonetheless greet us affectionately when we get home (well—*most* will). Their mealtimes have been transformed by today's pet-food industry from a chore into a picnic. They remain unobtrusive most of the time, yet seem delighted to receive our affection. In a word, they are *convenient*.

Yet despite their apparently effortless transformation into urban sophisticates, however, cats still have three out of four feet firmly planted in their wild origins. The dog's mind has been radically altered from that of its ancestor, the gray wolf; cats, on the other hand, still think like wild hunters. Within a couple of generations, cats can revert back to the independent way of life that was the exclusive preserve of their predecessors some ten thousand years ago. Even today, many millions of cats worldwide are not pets but feral scavengers and hunters, living alongside people but inherently distrustful of them. Due to the astonishing flexibility with which kittens learn the difference between friend and foe, cats can move between these dramatically different lifestyles within a generation, and the offspring of a feral mother and feral father can become indistinguishable from a cat descended from generations of pets. A pet that is abandoned by its owner and cannot find another may turn to scavenging; a generation or two on, and its descendants will be indistinguishable from the thousands of feral cats that live shadowy existences in our cities.

As cats become more popular and ever more numerous, those who revile them continue to raise their voices, but with more venom now than for several centuries before. Cats have never shared the "unclean" tag foisted on the dog and the pig, but despite cats' superficially universal acceptance, a minority of people across all cultures finds cats disagreeable, and as many as one in twenty say that they find them repulsive.[2] When asked, few Westerners will admit that they don't like dogs: those who do usually turn out either not to like animals in general or can trace their aversion to a specific experience, perhaps being bitten in childhood.[3] Cat-phobia is more deeply seated and less widespread than common phobias of snakes and spiders—phobias that have a logical basis in helping the sufferer to avoid poisonous varieties—but is just as powerful an experience for those who suffer from it.[4] Cat-phobics were likely at the forefront of the religious persecution that led to the killing of millions of cats in medieval Europe, and cat-phobia was likely just as common then as it is today. Thus, there can be no guarantee that the cat's popularity will last. Indeed, without our intervention, the twentieth century may turn out to have been the cat's golden age.

Today, spurred on by confessed cat-haters, the cat is coming under attack on the specific grounds that it is a wanton and unnecessary killer of "innocent" wildlife. These voices are most loudly raised in the Antipodes, but are becoming increasingly strident in the UK and the United States. The anti-cat lobby, at its most extreme, demands that cats no longer be allowed to hunt, that pet cats be kept indoors, and that feral cats should be exterminated. Owners of outdoor cats are vilified for supporting an animal that is portrayed as laying waste to the wildlife around their homes. Veterinarians who seek to manage the welfare of feral cats by neutering and vaccinating them, and then returning them to their original territory, have come under attack from within their own profession, with some colleagues charging that this constitutes (illegal) abandonment that benefits neither the cat nor the adjacent wildlife.[5]

Both sides in this debate admit that cats are "natural" hunters, but cannot agree on how this behavior might be managed. In parts of Australia and New Zealand, cats are defined as "alien" predators introduced from the Northern hemisphere, and are banned from some areas and subject to curfews or compulsory microchipping in others. Even in places where cats have lived alongside native wildlife for hundreds of years, such as in the United States and the UK, their increasing popularity as pets has prompted a vocal minority to press for similar restrictions. Cat owners point to a lack of scientific evidence that pet cats contribute significantly to a population decline of any wild bird or mammal, which are caused instead mainly by the recent proliferation of other pressures on wildlife, such as loss of habitat. Consequently, any restrictions imposed on pet cats are unlikely to result in a resurgence of the species that they supposedly threaten.

Cats themselves are of course unaware that we no longer value their hunting prowess. Insofar as they are concerned, the greatest threat to their subjective well-being comes not from people, but instead from other cats. In the same way that cats are not born to love people—this is something they have to learn when they are kittens—they do not automatically love other cats; indeed, their default position is to be suspicious, even fearful, of every cat they meet. Unlike the highly sociable wolves that were forebears to modern dogs, the ancestors of cats were both solitary and territorial. As cats began their association with humankind some 10,000 years ago, their tolerance for one another must have been forced to improve so that they could live at the higher densities that man's provision of food for them—at first accidental, then deliberate—allowed.

Cats have yet to evolve the optimistic enthusiasm for contact with their own kind that characterizes dogs. As a result, many cats spend their lives trying to avoid contact with one another. All the while, their owners inadvertently compel them to live with cats they have no reason to trust—whether the neighbors' cats, or the second cat obtained by the owner to "keep him company." As their popularity increases, so inevitably does the number of cats that each cat is forced into contact with, thereby increasing the tensions that each experiences. Finding it ever harder to avoid social conflict, many cats find it nearly impossible to relax; the stress they experience affects their behavior and even their health.

The well-being of many pet cats falls short of what it should be—perhaps because their welfare does not grab headlines in the way that dogs' welfare does, or perhaps because they tend to suffer in silence. In 2011, a UK veterinary charity estimated that the average pet cat's physical and social environment scored only 64 percent, with households that owned more than one cat scoring even lower. Owners' understanding of cat behavior scored little better, at 66 percent.[6] Without a doubt, if cat owners understood more about what makes their cats tick, many cats could live much happier lives.

Faced with such pressures, cats need not our immediate emotional reactions—irrespective of whether we find them endearing or not—but instead a better understanding of what they want from us. Dogs are expressive; their wagging tails and bouncy greetings tell us in no uncertain terms when they are happy, and they do not hesitate to let us know when they are distressed. Cats, on the other hand, are undemonstrative; they keep feelings to themselves and rarely tell us what they need, beyond asking for food when they're hungry. Even purring, long assumed to be an unequivocal sign of contentment, is now known to have more complex significance. Dogs certainly benefit from the knowledge of their true natures that can come only from science, but for cats this comprehension is essential, for they rarely communicate their problems to us until these have become too much to bear. Most of all, cats require our assistance when, as happens far too often, their social lives run awry.

Cats desperately need the kind of research from which dogs have benefited, but unfortunately feline science has not seen the explosion of activity that has recently occurred in canine science. Cats have simply not grabbed the attention of scientists as dogs have. However, the past two decades have provided significant advances, profoundly affecting scientists' interpretations of how cats view the world, and what makes them "tick." These new insights form the core of this book, giving us the first indications on how to help cats adjust to the many demands we now put on them.

Cats have adapted to live alongside people while retaining much of their wild behavior. Apart from the minority that belong to a breed, cats are not humankind's creation in the sense that dogs are; rather, they have coevolved with us, molding themselves into two niches that we have unintentionally provided for them. The first role for cats in human society was that of pest controller: some 10,000 years ago, wild cats moved in to exploit the concentrations of rodents provided by our first granaries, and adapted themselves to hunting there in preference to the surrounding countryside. Realizing how beneficial this was—cats, after all, had no interest in eating grain and plant foods themselves—people must have begun to encourage cats to stay by making available their occasional surpluses of animal products, such as milk and offal. The cats' second role, which undoubtedly followed hard on the heels of the first but whose origins are lost in antiquity, is that of companion. The first good evidence that we have for pet cats comes from Egypt some 4,000 years ago, but women and children in particular may have adopted kittens as pets long before this.

As far as humans are concerned, these dual roles of pest controller and companion have ceased to go hand-in-hand. Although we treasured cats until recently for their prowess as hunters, few owners today express delight when their cat deposits a dead mouse on their kitchen floor.

Cats carry the legacy of their primal pasts, and much of their behavior still reflects their wild instincts. To understand why a cat behaves as it does, we must understand where it came from and the influences that have molded it into what it is today. Therefore, the first three chapters of this book chart the cat's evolution from wild, solitary hunter to high-rise apartment-dweller. Unlike dogs, only a small minority of cats has ever been intentionally bred by people—and furthermore, when there has been deliberate breeding, it has been exclusively for appearance. No one has bred cats to guard houses, to herd livestock, or to accompany or assist hunters. Instead, cats have evolved to fill a niche brought about by the development of agriculture, from its beginnings in the harvest and storage of wild grains, to today's mechanized agribusiness.

Of course, when the cat first infiltrated our settlements many thousands of years ago, its other qualities did not go unnoticed. Its appealing features, its childlike face and eyes, the softness of its fur, and, crucially, its ability to learn how to become affectionate toward us, led to its adoption as a pet. Subsequently, humankind's passion for symbolism and mysticism elevated the cat to iconic status. Popular attitudes toward cats have been profoundly influenced by such connotations: extreme religious views toward cats have affected not only how they were treated, but their very biology—both how they behave and how they look.

Cats have changed to live alongside humans, but we have very different ways of gathering information about and thereby interpreting the physical world that we share. Chapters 4 through 6 examine those differences: humans and cats are both mammals, but our senses and brains work in different ways. Cat owners often underappreciate these differences: we have a natural tendency to assume that animals perceive the world around us in the same way as we do. Moreover, even in today's world of rationality and science, we still treat the world as if it were sentient, attributing intention to the weather, the earth, and the movements of the stars in the sky. How easy, therefore, it is to fall into the trap of thinking that because cats are communicative and affectionate, they must be, more or less, little furry humans.

Science, however, reveals that cats are anything but. Beginning with the way that every kitten constructs its own version of the world, with consequences that will last its entire lifetime, this part of the book describes how the cat gathers information about its surroundings, especially the way it uses its hypersensitive sense of smell; how its brain interprets and uses that information; and how its emotions guide its responses to opportunities and challenges alike. In scientific circles, it has only recently become acceptable to talk about animal emotions, and one school of thought still maintains that emotions are a byproduct of consciousness, meaning that no animals except humans and possibly a few primates can possibly possess them. However, common sense dictates that if an animal that shares our basic brain structure and hormone systems looks frightened, it must be experiencing something very like fear—probably not in quite the same way that we experience it, but fear nonetheless.

Most (but not everything) of what biology has revealed about the cat's world fits the idea that cats have evolved as predators first and foremost. Cats are social animals too; otherwise, they could never have become pets as well as hunters. The demands of domestication—first of all, the need to cohabitate with other cats in human settlements, and then the benefits of forming affectionate bonds with people—have extended cats' social repertoires out of all recognition compared to those of their wild ancestors. Chapters 7 through 9 explore these social connections in detail: how cats conceive of and interact with other cats and with people, and why two cats may react very differently in the same situation. In other words, we will examine the science of cat "personality."

The book concludes with an examination of the cat's current place in the world, and how this might evolve in the coming decades. Cats are under pressure from many different interests, some well-meaning and others antagonistic. Pedigree cats are still in a minority, and those who breed them are in a position to avoid the practices that have so adversely affected the welfare of pedigree dogs over the past few decades.[7] However, the growing fashion for hybrids between domestic cats and other, wild, species of cat, resulting in "breeds" such as the Bengal, can have unintended consequences. We must also ask whether the cat is being inadvertently and subtly altered by those who hold cat welfare closest to their hearts. Paradoxically, the drive to neuter as many cats as possible, with its laudable aim of reducing the suffering of unwanted kittens, may be gradually eliminating the characteristics of the very cats best suited to living in harmony with humankind: many of the cats that avoid neutering are those that are most suspicious of people and the best at hunting. The friendliest, most docile cats are nowadays neutered before leaving any descendants, while the wildest, meanest ferals are likely to escape the attention of cat rescuers and breed at will, thus pushing the cat's evolution away from, rather than toward, better integration with human society.

We are in danger of demanding more from our cats than they can deliver. We expect that an animal that has been our pest controller of choice for thousands of years should now give up that lifestyle because we have begun to find its consequences distasteful or unacceptable. We also expect that we should be free to choose our cat's companions and neighbors without regard for their origins as solitary, territorial animals. Somehow, we presume that because dogs can be flexible in their choice of canine companions, cats will be equally tolerant of whatever relationships we expect them to develop, purely for our convenience.

Until about twenty or thirty years ago, cats kept pace with human demands, but they are now struggling to adapt to our expectations, especially that they should no longer hunt, and no longer desire to roam away from home. In contrast to almost every other domestic animal, whose breeding has been strictly controlled for many generations past, the cat's transition from wild to domestic has—with the exception of pedigree cats—been driven by natural selection. Cats essentially evolved to fit opportunities that we provided. We allowed them to find their own mates, and those kittens that were best suited to living alongside humans, in whatever capacity was required of them at the time, were the most likely to thrive and produce the next generation.

Evolution is not going to produce a cat that has no urge to hunt and that is as socially tolerant as a dog—at least not within a timescale that will be acceptable to the cat's detractors. Ten thousand years of natural selection has provided the cat with enough flexibility to fend for itself when, from time to time, its compact with man breaks down, but not enough to cope with a demand that has grown from nowhere in just a few years. Even for such a prolific breeder as the cat, natural selection would take many generations to move even a token step in this direction. Only deliberate, carefully considered breeding can produce cats that are well suited to the demands of tomorrow's owners, and that will be more acceptable to cat haters.

We can do much to improve the lot of cats today. Better socialization of kittens, better understanding of what environments cats really need, more deliberate intervention in teaching cats to cope with situations that they find distressing—all these can help cats to adjust to the demands we now make of them, and can also deepen the bond between cat and owner.

Cats are in many ways the ideal pet for the twenty-first century, but will they be able to adapt to the twenty-second? If they are to continue to remain in our affections—and the persecution they have received in the past indicates that this is hardly a given—then some consensus must emerge among cat welfare charities, conservationists, and cat fanciers on how to produce a type of cat that checks all the boxes. These changes must be guided by science. Initially, the way forward will be for cat owners and the general public alike to understand better where cats came from and why they behave as they do. At the same time, owners can rehabilitate the cat's fraying reputation by learning how to channel their cat's behavior, not only to discourage them from hunting, but also to make them happier in themselves. In the longer term, the emerging science of behavioral genetics—the mechanics of how behavior and "personality" are inherited—will allow us to breed cats that can better adapt to an ever-more crowded world.

As history shows, cats can fend for themselves in many ways. However, they cannot face what society now demands of them without human assistance. Our understanding of cats must start with a healthy respect for their essential natures.

CHAPTER 1

The Cat at the Threshold

Pet cats are now a global phenomenon, but how they transformed themselves from wild to domestic is still a mystery. Most of the animals around us were domesticated for prosaic, practical reasons. Cows, sheep, and goats provide meat, milk, and hides. Pigs provide meat; chickens, meat and eggs. Dogs, our second favorite pet, continue to provide humans with many benefits beside companionship: help with hunting, herding, guarding, tracking, and trailing, to name but a few. Cats are not nearly as useful as any of these; even their traditional reputation as rodent controllers may be somewhat exaggerated, even though, historically, this was their obvious function as far as humankind was concerned. Therefore, in contrast to the dog, we have no easy answers as to how the cat has insinuated itself so effectively into human culture. Our search for explanations starts some ten millennia ago, when cats probably first arrived at our doorsteps.

Conventional accounts of the domestication of the cat, based on archaeological and historical records, propose that they first lived in human homes in Egypt about 3,500 years ago. This theory, however, has recently been challenged by new evidence coming from the field of molecular biology. Examination of differences between the DNA of today's domestic and wild cats has dated their origins much earlier, anywhere between ten thousand and fifteen thousand years ago (8,000 and 13,000 years BCE). We can safely discount the earliest date in this range—anything earlier than about 15,000 years ago makes little sense in terms of the evolution of our own species, since it is unlikely that stone-age hunter-gatherers would have had the need or resources to keep cats. The minimum estimate, 10,000 years, presumes that domestic cats are derived from several wild ancestors that came from several different locations in the Middle East. In other words, the domestication of the cat happened in several widely separated places, either roughly contemporaneously or over a longer period of time. Even if we assume that cats started to become domesticated around 8,000 years BCE, this leaves us with a 6,500-year interval before the first historical records of domestic cats appear in Egypt. So far, few scientists of any kind have studied this first—and longest—phase in the partnership between human and cat.

The archaeological record for this period, such as it is, is not very illuminating. Cats' teeth and fragments of bones dating between 7,000 and 6,000 BCE have been excavated around the Palestinian city of Jericho and elsewhere in the Fertile Crescent, the "cradle of civilization" that extends from Iraq through Jordan and Syria to the eastern shores of the Mediterranean and Egypt. However, these fragments are uncommon; moreover, they could well have come from wild cats, perhaps killed for their pelts. Rock paintings and statuettes of catlike animals from the following millennia, discovered in what is now Israel and Jordan, could conceivably depict domesticated cats; however, these cats are not depicted in domestic settings, so they may well be representations of wild cats, possibly even big cats. Yet even if we assume that these pieces of evidence all refer to early forms of the domestic cat, their very rarity still must be explained. By 8,000 BCE, humankind's relationship with the domestic dog had already progressed to the extent that dogs were routinely buried alongside their masters in several parts of Asia, Europe, and North America, whereas burials of cats first became common in Egypt around 1,000 BCE.[1] If cats were indeed domesticated pets during this time, we should have far more tangible evidence of that relationship than has been uncovered.

Our best clues on how the partnership between man and cat began come not from the Fertile Crescent, but instead from Cyprus. Cyprus is one of the few Mediterranean islands that have never been joined to the mainland, even when that sea was at its lowest level. Consequently, its animal population has had to migrate there by flying or swimming—that is, until humans started to travel there in primitive boats some 12,000 years ago. At that point, the Eastern Mediterranean had no domesticated animals, with the likely exception of some early dogs, so the animals that made the crossing with those first human settlers must have been either individually tamed wild animals or inadvertent hitchhikers. Therefore, while we cannot possibly tell whether ancient remains of cats on the mainland are from wild, tame, or domesticated animals, cats could clearly have reached Cyprus only by being deliberately transported there by humans—assuming, as we safely may, that cats of that era were as averse to swimming in the ocean as today's cats are. Any remains of cats found there must be those of semidomesticated or at least captive animals, or their descendants.

On Cyprus, the earliest remains of cats coincide with and are found within the first permanent human settlements, some 7,500 years BCE, making it highly likely that they were deliberately transported there. Cats are too large and conspicuous to have been accidentally transported across the Mediterranean in the small boats of the time: we know very little about seagoing boats from that period, but they were likely too small to conceal a stowaway cat. Moreover, we have no evidence for cats living away from human habitations on Cyprus for another 3,000 years. The most likely scenario, then, is that the earliest settlers of Cyprus brought with them wildcats that they had captured and tamed on the mainland. It is implausible that they were the only people to have thought of taming wildcats, so capturing cats and taming them were likely already an established practice in the Eastern Mediterranean. Confirming this, we also have evidence for prehistoric importations of tamed cats to other large Mediterranean islands, such as Crete, Sardinia, and Majorca.

The most likely reason for taming wildcats is also evident from the first settlements on Cyprus. Right from the outset, these habitations, like their counterparts of the time on the mainland, became infested by house mice. Presumably these unwanted mice *were* stowaways, accidentally transported across the Mediterranean in sacks of food or seed corn. The most likely scenario, therefore, is that as soon as mice became established on Cyprus, the colonizers imported tame or semidomesticated cats to keep them under control. This might have been ten years or a hundred years after the first settlements were established—the archaeological record cannot reveal such small differences. If this is correct, it suggests that the practice of taming cats to control mice was already entrenched on the mainland as long as 10,000 years ago. No firm evidence for this is ever likely to be found, because the ubiquitous presence of wildcats there makes it impossible to tell whether the remains of a cat, even if found within a settlement, are those of a truly wild cat that had died or been killed when hunting there, or of a cat that had lived there for most or all of its life.

Whatever its exact origins, the tradition of taming wildcats to control vermin continued into modern times in parts of Africa where domestic cats are scarce and wildcats easy to obtain. While traveling the White Nile in 1869, the German botanist-explorer Georg Schweinfurth found that his boxes of botanical specimens were being invaded by rodents during the night. He recalled:

> One of the commonest animals hereabouts was the wild cat of the steppes. Although the natives do not breed them as domestic animals, yet they catch them separately when they are quite young and find no difficulty in reconciling them to a life about their huts and enclosures, where they grow up and wage their natural warfare against the rats. I procured several of these cats, which, after they had been kept tied up for several days, seemed to lose a considerable measure of their ferocity and to adapt themselves to an indoor existence so as to approach in many ways to the habits of the common cat. By night I attached them to my parcels, which were otherwise in jeopardy, and by this means I could go to bed without further fear of any depredations from the rats.[2]

Like Schweinfurth, those much earlier explorers who first brought wildcats to Cyprus would almost certainly have found that they had to keep the cats tethered. If allowed to run free, the cats would have quickly escaped and wreaked havoc on the native fauna, which up to that point would have had no experience with a predator as formidable as a cat. We know that this is what eventually happened. Several centuries after human settlement, cats indistinguishable from wildcats spread throughout Cyprus and remained there for several thousand years.[3] Most likely, only the cats that were confined to the grain stores would have stayed there to help the early settlers to rid those stores of pests: The others would have left to exploit the local wildlife. The descendants of these escapees may have been captured and even eaten from time to time, since broken cat bones have been found at several other Neolithic sites on Cyprus, as well as those of other predators such as foxes and even domestic dogs.

The practice of taming wildcats to control vermin was probably prompted by the emergence of a new pest in the early granaries, the house mouse (*Mus musculus*); indeed, the histories of these two animals are inextricably interwoven. The house mouse is one of more than thirty species of mouse found worldwide, but the only one that has adapted to living alongside humans and exploiting our food.

House mice have their origins in a wild species from somewhere in northern India that was in existence possibly as long as a million years ago, certainly well before the evolution of humankind. From there they spread both east and west, feeding on wild grains, until some reached the Fertile Crescent, where they eventually encountered the earliest stores of harvested grain: mouse teeth have been found among stored grain dating back 11,000 years in Israel, and a 9,500-year-old stone pendant carved in the shape of a mouse's head was found in Syria. Thus began an association with humankind that continues to the present day. Humans not only provided an abundance of food that mice could exploit, but our buildings also provided both warm, dry places to build nests and protection from predators such as wildcats. Mice that could adapt to these living conditions thrived, while those that could not died out: today's house mice rarely breed successfully away from human habitation, especially where there are wild competitors, such as wood mice.

Humans also provided house mice with a way to colonize new areas. Mice from the southeastern part of the Fertile Crescent, what is now Syria and northern Iraq, were accidentally transported, presumably in grain being traded between communities, throughout the Near East, up to the eastern shores of the Mediterranean, and then to nearby islands such as Cyprus.

The first culture to be bedeviled by house mice was that of the Natufians, who by extension are the people most likely to have initiated the cat's long journey into our homes. The Natufians inhabited the area that now comprises Israel-Palestine, Jordan, southwestern Syria, and southern Lebanon from about 11,000 to 8,000 BCE. Widely regarded as the inventors of agriculture, they were initially hunter-gatherers like other inhabitants of the region; soon, however, they began to specialize in harvesting the wild cereals that grew abundantly all around them, in a region that was significantly more productive then than it is today. To do this, the Natufians invented the sickle. Sickle blades found at Natufian settlements still show the glossy surfaces that could have been produced only from scything through the abrasive stems of wild grains—wheat, barley, and rye.

The early Natufians lived in small villages; their houses were partly belowground, partly above, with walls and floors of stone and brushwood roofs. Until about 10,800 BCE, they rarely planted cereals deliberately, but over the next 1,300 years a rapid change in climate, known as the Younger Dryas, brought about a significant intensification in field-clearing, planting, and cultivation. As the amounts of harvested grain increased, so did the need for storage. The Natufians and their successors probably used storage pits built out of mudbrick and constructed like miniature versions of their houses. It was probably this invention that triggered the self-domestication of the house mouse, which, moving into this rich and novel environment, thereby became humankind's first mammalian pest species.

As the numbers of mice grew, they must have attracted the attentions of their natural predators, including foxes, jackals, birds of prey, the Natufians' domestic dogs, and, of course, wildcats. Wildcats had two advantages that set them apart from other predators of mice: they were both agile and nocturnal, well adapted to hunting in the near-darkness when the mice became active. However, if these wildcats had been as frightened of man as their modern counterparts are, it is difficult to imagine how they would have exploited this new rich source of food. Almost certainly, therefore, the wildcats in the region inhabited by the Natufians were less wary than those of today.

We have no evidence that the Natufians deliberately domesticated the cat. Like the mouse before it, the cat simply arrived to exploit a new resource that had been created by the beginnings of agriculture. As Natufian agriculture became more complex, involving both an increasing array of crops and the domestication of animals such as sheep and goats, and as agriculture extended to other regions and cultures, so the opportunities available for cats multiplied. These were not pet cats as we know them today; rather, the cats that exploited these concentrations of mice would have been more like today's urban foxes—capable of adapting to a human environment, but still retaining their essential wildness. Domestication was to come much later.

We know surprisingly little about the wildcats of the Fertile Crescent and surrounding areas (see box on page 8, "The Evolution of the Cats"). The archaeological record indicates that 10,000 years ago several species lived in the region, all of which would have been attracted by concentrations of mice. We know that later on, the ancient Egyptians kept tame jungle cats, *Felis chaus*, in considerable numbers; jungle cats, though, are substantially heavier than wildcats, weighing between ten and twenty pounds, and large enough to kill young gazelle and chital. Although their normal diet includes rodents, they may have been too obtrusive to get regular access to granaries. Alternatively, they may simply have been temperamentally unsuited to living alongside man. We do have evidence that the Egyptians tried to tame and even train them as rodent controllers, but apparently without any lasting success.

The Evolution of the Cats

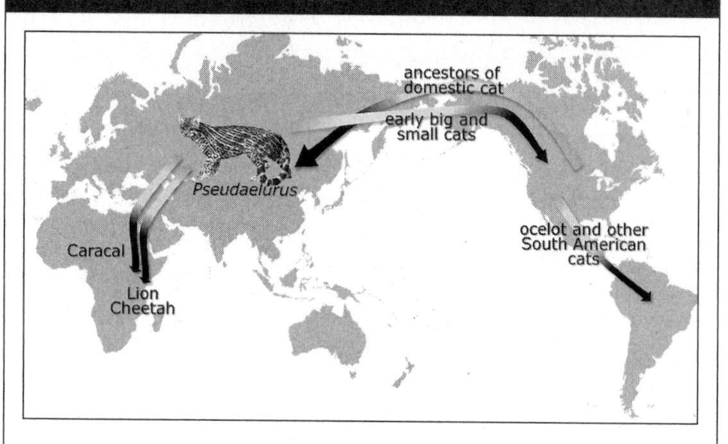

Every member of the cat family, from the noble lion to the tiny black-footed cat, can trace its ancestry back to a medium-sized catlike animal, *Pseudaelurus*, that roamed the steppes of central Asia some 11 million years ago. *Pseudaelurus* eventually went extinct, but not before unusually low sea levels had allowed it to migrate across what is now the Red Sea into Africa, where it evolved into several medium-sized cats, including those we know today as the caracal and the serval. Other *Pseudaelurus* traveled east across the Bering land bridge into North America, where they eventually evolved into the bobcat, lynx, and puma. Some 2 to 3 million years ago, following the formation of the Panama isthmus, the first cats crossed into South America; here they evolved in isolation, forming several species not found anywhere else, including the ocelot and Geoffroy's cat. The big cats—lions, tigers, jaguars, and leopards—evolved in Asia and then spread into both Europe and North America, their present-day distributions but a tiny relict of where they used to roam a few million years ago. Remarkably, the distant ancestors of today's domestic cats seem to have evolved in North America about 8 million years ago, and then migrated back into Asia some 2 million years later. About 3 million years ago, these began to evolve into the species we know today, including the wildcat, the sand cat, and the jungle cat; a separate Asian lineage, including Pallas's cat and the fishing cat, also began to diverge at about this time.[4]

Jungle cat

Coeval with them were sand cats, *Felis margarita*, large-eared nocturnal animals that hunt by night, using their acute hearing. They are, moreover, comparatively unafraid of humans, and hence might be thought good candidates for taming and domestication. However, they are made for life in deserts—the pads on their feet are covered in thick fur to protect them from the hot sand—so few would have found themselves near the first stores of grains: the Natufians generally built their villages in wooded areas.

Sand cat

As civilization spread eastward through Asia, so it would have come into contact with other cat species. At Chanhudaro, a town built by the Harappan civilization close to the Indus River in what is now Pakistan, archaeologists found a 5,000-year-old mudbrick imprinted with a cat's foot, overlapped by that of a dog. As the newly made brick was drying in the sun, the cat appears to have run across it, closely followed by a dog, possibly in hot pursuit. The footprint is larger than that of a domestic cat, and its webbed feet and extended claws identify it as a fishing cat, *Felis viverrina*, found today from the Indus basin eastward and south to Sumatra in Indonesia (though not in the Fertile Crescent). As its name implies, the fishing cat is a strong swimmer and specializes in catching fish and aquatic birds. Although it will also take small rodents, it is difficult to see how it would switch to a diet consisting predominantly of mice, so it too is an unlikely candidate for domestication.

Farther afield, we know of at least two other species of cat that came in out of the wild to prey on the vermin that plagued humankind's food stores. In Central Asia and ancient China, the local wild-

Manul

Jaguarundi

cat, the manul (or Pallas's cat, named after the German naturalist who first categorized it) was occasionally even tamed and deliberately kept as a rodent controller. The manul has the shaggiest coat of any member of the cat family, so long that its hair almost completely obscures its ears. In pre-Columbian Central America, meanwhile, an otter-like cat, the jaguarundi, was probably also kept as a semi-tame pest controller. None of these species have ever become fully domesticated, neither are any of them included in the direct ancestry of today's house cats.

Out of all these various wild cats, only one was successfully domesticated. This honor goes to the Arabian wildcat *Felis silvestris lybica*, as confirmed by their DNA.[5] In the past, both scientists and cat-fanciers have suggested that certain breeds within the domestic cat family are hybrids with other species—for example, the Persian's fluffy feet are superficially similar to the sand cat's, and its fine coat is somewhat like that of a manul. However, the DNA of all domestic cats—random-bred, Siamese, or Persian—shows no trace of these other species, or indeed any other admixture. Somehow, the Arabian wildcat alone was able to inveigle itself into human society, outcompeting all its rivals, and eventually spreading throughout the world. Although the qualities that gave it this edge are not easy to pin down, they probably occurred in combination only in the wildcats of the Middle East.

The wildcat *Felis silvestris* is currently found throughout Europe, Africa, and central Asia, as well as western Asia, the area where it probably first evolved. Like many predators, such as the wolf, it is now found only in isolated and generally remote areas where it can avoid persecution from man. This has not always been the case. Five thousand years ago, wildcats were evidently regarded as delicacies in some areas; the rubbish pits left by the "lake dwellers" of Germany and Switzerland contain many wildcat bones.[6] The cats must have been abundant at the time; otherwise, they could hardly have been trapped in such large numbers. Over the centuries they became less common, displaced by the felling of their forest habitat for agriculture, and forced farther into the woods by development and loss of habitat. The invention of firearms led to wildcats being hunted to extinction in many areas. During the nineteenth century, various European countries, including the UK, Germany, and Switzerland, classified them as vermin, due to the harm they supposedly caused both wildlife and livestock.[7] Only recently, due to the establishment of wildlife reserves and a more informed attitude to the important role that predators play in stabilizing ecosystems, are wildcats returning to areas such as Bavaria, where they have not been seen for hundreds of years.

The wildcat is now divided into four subspecies or races. These are the European forest cat *Felis silvestris silvestris*, the Arabian wildcat *Felis silvestris lybica*, the Southern African wildcat *Felis silvestris cafra*, and the Indian desert cat *Felis silvestris ornata*.[8] All these cats are rather similar in appearance, and all are capable of interbreeding where their ranges overlap. A possible fifth subspecies is the very rare Chinese desert cat *Felis bieti*, which according to its DNA split off from the main wildcat lineage about a quarter of a million years ago. It's possible that these cats actually form a separate species, as no hybrids are known to exist, but they live in such a small and inaccessible region—part of the Chinese province of Sichuan—that this may be due to lack of opportunity rather than physical impossibility.

Wildcats from different parts of the world differ markedly in how easily they can be tamed. Domestication, moreover, can start only

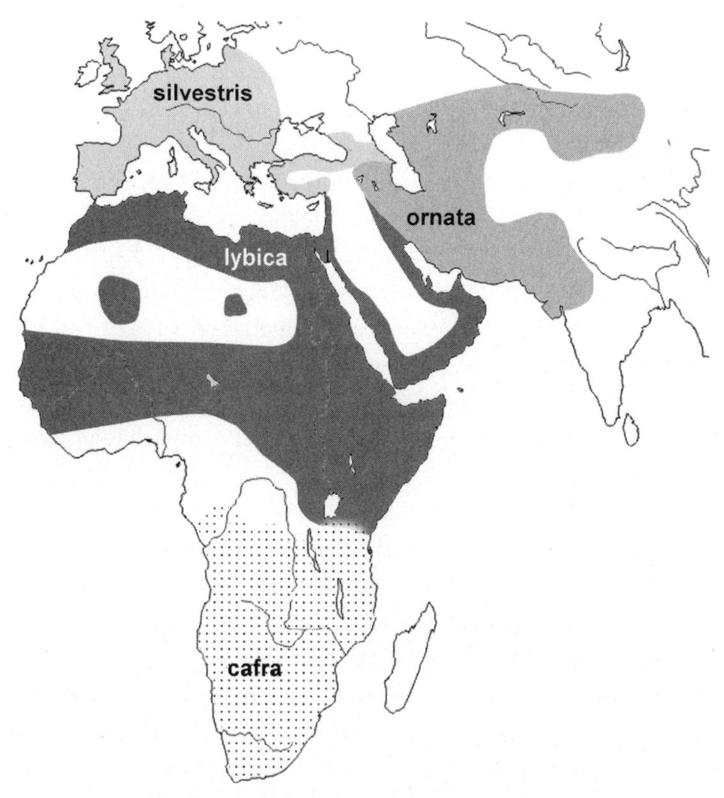

The distribution of the subspecies of wildcat

with animals that are already tame enough to raise their young in the proximity of people. Those offspring that are best suited to the company of humans and human environments are, perhaps unsurprisingly, more likely to stay and breed there than those that are not; the latter will most likely revert to the wild. Over several generations, this repeated "natural" selection will, even on its own, gradually change the genetic makeup of these animals so that they become better adapted to life alongside people. It is also likely that, at the same time, humans will intensify that selection, by feeding the more docile animals and driving away those prone to bite and scratch. This process cannot start without some genetic basis for tameness existing beforehand, and in the case of wildcats, this is far from evenly distributed. Today, some parts of the world have little raw material for domestication, while others seem more promising.

We know, for example, that the four subspecies of wildcat differ in how easy they are to tame. The European forest cat is larger and thicker-set than a typical domestic cat, and has a characteristic short tail with a blunt, black tip. This aside, it looks from a distance much like a domestic striped tabby—a distant glimpse is all that most people are likely ever to get, however, for it is among the wildest of animals. This is largely due to its genetics, and not the way it is raised: those few people who have tried to produce tame forest cats have met with precious little but rejection. In 1936, natural and wildlife photographer Frances Pitt wrote:

> It has long been stated that the European wildcat is untameable. There was a time when I did not believe this. . . . My optimism was daunted when I made acquaintance with Beelzebina, Princess of Devils. She came from the Highlands of Scotland, a half-grown kitten that spat and scratched in fiercest resentment. Her pale green eyes glared savage hatred at human-beings, and all attempts to establish friendly relations with her failed. She grew less afraid, but as her timidity departed, her savagery increased.[9]

Pitt then went on to obtain an even younger male kitten, in the hope that Beelzebina had been too old to be socialized when first found. That she named this new kitten Satan perhaps suggests how difficult he was to handle from the outset. As he grew stronger and more confident, he became impossible to touch; he would take food from the hand, but would spit and growl while doing so, and then quickly back away. However, he was not pathologically aggressive—he just hated people. While he was still young, Pitt introduced him to a female domestic kitten, Beauty, toward whom he was "all gentleness and devotion." When she was let out of the cage in which he had to be kept, "this distressed him sorely. He rent the air with harsh cries, for his voice, though loud, was not lovely." Beauty and Satan produced several litters of kittens, all of which had the characteristic appearance of forest cats. Some, despite being handled from an early age, grew to be as savage as their father; others were more sociable toward Pitt and her parents, though all remained very wary of unfamiliar people. Pitt's experiences with Scottish wildcats seem to be typical: Mike Tomkies, the "Wilderness Man," was also unable to socialize his two hand-raised wildcat sisters, Cleo and Patra, which he kept at his remote cottage on the shores of a Scottish loch.[10]

We know little about the Indian desert cat, but it is reputedly difficult to tame. This subspecies is found to the south and east of the Caspian Sea, southward through Pakistan and into the northwestern Indian states of Gujarat, Rajasthan, and the Punjab, and eastward through Kazakhstan into Mongolia. Its coat is usually paler than that of the other wildcats, and is blotchy rather than tabby in pattern. Like other wildcats, it will occasionally base itself near farms, attracted by the concentration of rodents, but it has never taken the next step to domestication—acceptance of humans. We have records from Harappa of tamed caracals, a medium-sized, long-limbed cat with characteristic tufted ears, and jungle cats, to add to the fishing cat that left its footprint there; but we find no indications of any Indian desert cats. For a long time biologists and cat fanciers alike thought that Siamese cats could be a mixture of domestic and Indian desert cat, the progeny of interbreeding between early domestic cats and local wildcats somewhere around the Indus valley. However, scientists have not found the characteristic DNA signature of the Indian desert cat in any examples of the Siamese and related breeds, which instead are ultimately derived from the wildcats of the Middle East or Egypt—there are no *silvestris* wildcats in Southeast Asia, so the original Siamese cats must have arrived from the west as fully domesticated animals.

The wildcats of South Africa and Namibia—"caffre cats"—are likewise genetically distinctive. They migrated south from the original wildcat population in northern Africa about 175,000 years ago, around the same time as the ancestors of the Indian desert cat migrated east. It is unclear where the boundary between the Southern African and Arabian wildcats lies—no wildcat's DNA has yet been characterized from any part of Africa except Namibia and the Republic of South Africa. The wildcats of Nigeria are shy, aggressive, and difficult to tame; those of Uganda are sometimes more tolerant of people, but many do not look like typical wildcats—which in that area have distinctive red-brown backs to their ears—and are probably hybrids, their domesticated genes accounting for their friendly behavior. Most of the street cats in the same region show signs of some wildcat in their ancestry, so the distinction between wildcat, street cat, and random-bred pet is blurred in many parts of Africa.

The wildcats of Zimbabwe—presumably belonging to the Southern African subspecies—are a case in point. In the 1960s, naturalist and museum director Reay Smithers kept two hand-reared female wildcats, Goro and Komani, at his home in what was then Southern Rhodesia.[11] Both were tame enough to be let out of their pens, though only one at a time, since they would fight whenever they met. Once, Komani disappeared for four months, finally reappearing one evening in the beam of Smithers's flashlight: "I called my wife, to whom she is particularly attached, and we sat down while she softly called the cat's name. It must have taken a quarter of an hour before Komani suddenly responded and came to her. The reunion was most moving, Komani going into transports of purring and rubbing herself against my wife's legs."

Such behavior is identical to that of a pet cat being reunited with its owner, and the similarities with pet cats did not end there. Both Goro and Komani were affectionate toward Smithers's dogs, rubbing themselves on their legs and curling up in front of the fire with them. Every day they demonstrated their affection for Smithers himself by an effusive display of typical pet cat behavior.

> These cats never do anything by halves; for instance when returning from their day out they are inclined to become super-affectionate. When this happens, one might as well give up what one is doing, for they will walk all over the paper you are writing on, rubbing themselves against your face or hands; or they will jump on your shoulder

> and insinuate themselves between your face and the book you are reading, roll on it, purring and stretching themselves, sometimes falling off in their enthusiasm and, in general, demanding your undivided attention.

This may be the behavior of a typical hand-reared "caffre cat," but it is more likely that Goro and Komani, while undoubtedly wildcats in terms of their markings and their hunting ability, nevertheless contained some DNA from interbreeding with pet cats somewhere in their ancestry. The extent of hybridization between wildcats and domestic cats in South Africa and Namibia was recently revealed by DNA sequences from twenty-four supposed wildcats, eight of which bore the telltale signs of partial descent from domestic cats. In a survey of zoos in the United States, the UK, and the Republic of South Africa, I found that ten out of twelve South African wildcats displayed affectionate behavior toward their keepers, and of these, two would regularly rub and lick them.[12] This kind of behavior strongly suggests that the latter were hybrids, while those that could not be handled at all were probably genuine wildcats. The eight that were moderately affectionate might have been either.

Hybridization between wildcats and domestic cats is not confined to Africa. In one study, five out of seven wildcats collected in Mongolia carried traces of domestic cat DNA; only two were "pure" Indian desert cat. In my survey of cats in zoos, I found that out of a dozen cats of this subspecies kept in captivity, only three had ever spontaneously approached their keepers, and only one had ever rubbed on its keeper's leg. From the proportions found in the DNA results, it seems highly likely that all of these were hybrids, even though they all looked like typical Indian desert cats. In the same study of wildcat DNA, almost a third of apparent "wildcats" sampled in France had some ancestry from domestic pets.[13] With the advent of DNA technology, it is easy to detect hybridization when the local wildcats are genetically distinct from domestic cats—as they are in southern Africa, central Asia, and western Europe alike. Defining what is wild and what is a hybrid is much more problematic in places where domestic cats and wild cats are genetically almost identical, as they are around the Fertile Crescent, home of the Arabian wildcat.

The Arabian wildcat *lybica* is not only the most similar to domestic cats, it is also probably the nearest living representative of the first *Felis silvestris*, all the other subspecies having evolved hundreds of thousands of years ago—a consequence of small numbers of animals migrating east, south, or west from the species' origin in the Middle East. The wildcats in Africa north of the Sahara are also probably *lybica*, but their DNA has not yet been tested to confirm this. Like all wildcats, the Arabian/North African wildcat has a "mackerel" striped tabby coat, varying in color from gray to brown—darkest in forest-dwelling animals, palest in those that live on the edges of deserts. It is generally larger and leaner than a typical domestic cat, and both its tail and legs are especially long; indeed, the front legs are so long that when it sits, its posture is characteristically upright, as depicted by the Ancient Egyptians in statues of the cat goddess Bast. While generally nocturnal and therefore rarely seen, it is not particularly rare. Although it is widely claimed that the Arabian wildcat's kittens, if hand-raised, become affectionate toward people, most eyewitness accounts come from central or southern Africa, and therefore probably refer to *cafra* rather than *lybica*. The explorer Georg Schweinfurth procured his tame wildcats in what is now Southern Sudan, roughly where the ranges of *lybica* and *cafra* merge together, and the most northerly location in Africa to yield reliable accounts of tamable wildcats.

Very little is known of the behavior of genuine *lybica* wildcats, either in the Middle East or northeast Africa. In the 1990s, conservationist David Macdonald radio-collared six wildcats on the Thumamah reserve in central Saudi Arabia. All except one kept their distance from human activity: the sixth, however, "often wandered into the vicinity of the pigeon house [in Thumamah town] and would often be found sleeping with the domestic cats in the yard of one of the houses. On one occasion he was seen mating with a [domestic] cat."[14] Apart from showing just how easily hybridization between wild and domestic cats can occur, these and other observations shed little

The Arabian wildcat—*Felis lybica*

light on whether the wildcats in this part of the world might have been easy to tame, thousands of years ago.

Tracing the precise geographic origin of the domestic cat is therefore far from easy, for if the archaeological evidence is inconclusive, so is the most recent DNA evidence. The domestic cat's genetic footprint has spread throughout the world and, because it interbreeds so readily, it is now found almost universally in what are, to all outward appearances, "wildcats." This is true wherever wildcats have been investigated, from Scotland in the north, to Mongolia in the east, to the southern tip of Africa. Many of these apparent "wildcats" have DNA characteristic of domestic cats and must therefore be mainly or entirely descended from domestic cats that have gone feral. Others have a mixture—part wildcat DNA, part domestic. Of thirty-six wildcats sampled in France, twenty-three had "pure" wildcat DNA, eight were indistinguishable from domestic cats, and five were evidently a mixture of the two. The techniques used are sensitive enough only to detect the major contributions to each cat's ancestry: a cat with one domestic and fifteen wild great-great-grandparents, for example, would probably show up as "pure" wildcat.

Bearing all this in mind, very few entirely "pure" *Felis silvestris* wildcats are left anywhere in the world. At least a millennium of contact between wildcat and domestic cat—and four to ten times as long in the Middle East—means that there must be at least one hybrid in virtually every free-living cat's ancestry. At one extreme, some are domestic pets that have taken to the wild and happen to have the right "mackerel" coat, such that if they are hit by a vehicle or trapped in some remote area, they are labeled as wildcats—only a sample of their DNA gives away their true identity. At the other end of the spectrum, some wildcats' ancestry might extend back several hundred generations before a domestic type crops up in their otherwise "unblemished" family tree.

For conservationists anxious to preserve the wildcat in its pristine state, this is an inconvenient truth. In many places in Europe, wildcats are protected animals, and it is an offense to kill one deliberately: feral cats are not offered this status, and may even be treated as vermin. To be clear, a feral cat is a cat that is living wild but is descended from domestic cats: most are distinguishable from wildcats by their markings, which can be any color found in domestic cats. How is the law to operate if there is no hard-and-fast genetic distinction between ferals and wildcats? The best answer is probably the pragmatic one: if a cat looks like a wildcat and behaves like a wildcat—that is, it lives by hunting rather than scavenging—then it probably is a wildcat, or near enough to a wildcat to make little difference. Domestic cats, even those that grow up in the wild and have to fend for themselves, are rarely as skilled at hunting as genuine wildcats. Furthermore, we can now identify the purest wildcats from their DNA, sampled from just a few hairs: individual cats might therefore be given special protected status in the confidence that they are not just domestic "lookalikes."

This near-universal hybridity makes it difficult to pinpoint the origin of the domestic cat—difficult, but not necessarily impossible. Wherever they originated, all the wildcats in that one location should have domestic-type DNA. Following some 4,000 years of coexistence, and presumably interbreeding, each will carry different proportions of wild and domestic genes, but these will be indistinguishable (except for the fifteen to twenty genes, so far unidentified, that make cats easier or harder to socialize to people; these must by definition be different between domestic and wildcats).[15] Unfortunately, largely due to the current turmoil in the Middle East and North Africa, it is difficult to obtain sufficient samples of DNA from wildcats in the Fertile Crescent and northeast Africa to probe this hypothesis fully. The most comprehensive study done so far was able to include samples from only two colonies of cats in southern Israel, three individuals collected in Saudi Arabia, and one in the United Arab Emirates. There were no samples from Lebanon, Jordan, Syria, or Egypt, nor were any included from north Africa—so even the cats from Libya that give *lybica* its name are still not definitively classified.[16] Until there is more information about the DNA of cats from all these regions, it is impossible to use genetic information to say precisely where domestication began.

What the diversity of DNA among today's cats does suggest is that not just one but several populations of scavenging cats were domesticated. These several domestications may have been more or less contemporary, but it is more likely that they occurred hundreds or even thousands of years apart. We can be reasonably sure that none of those domestications took place in Europe, India, or Southern Africa; otherwise, we would find traces of the DNA of wildcats from those regions in modern domestic cats. But precisely where in western Asia and/or northeast Africa those transformations took place awaits further research.

Using the available data, we can posit a convincing scenario, which is that cats were first tamed in one location, probably in the Middle East, for rodent control. The most likely area is therefore that inhabited by the Natufians, but they were not the only early grain-grinding culture in that part of the world. Even earlier, approximately 15,000 years ago in what is now Sudan and southern Egypt, the Qadan culture lived in fixed settlements and harvested wild grains in large quantities. However, some 4,000 years later, following a series of devastating floods in the Nile valley, they were displaced by hunter-gatherers, meaning that if they had tamed their local wildcats to protect their grain stores, this practice may have died out as their culture was destroyed. During roughly the same period, but further north, in the Nile valley, the Mushabians are thought to have independently developed some of the technologies that eventually led to agriculture, including food storage and the cultivation of figs. Again, they might feasibly have tamed wildcats to protect their food stores. About 14,000 years ago some Mushabians left Egypt and moved northeast into the Sinai desert, where, mixing with the local Kebaran people, they became the Natufians.[17] These migratory Mushabians apparently lived as nomadic hunter-gatherers, but it is possible that, even if they had no need or capacity to bring tame cats with them, they nevertheless brought an oral tradition of the usefulness of cats that was absorbed into the Natufian culture.

Even if we give the Natufians sole credit for the first domestication of the cat, the genetic diversity of today's cats must have resulted from domestication in more than one location. Wildcats from any one area tend to be genetically similar because they are territorial animals and rarely migrate. Flow of genes between regions is a very slow process—slow, that is, until more recently, due to humankind's intervention. We know that domestic cats are perfectly able, indeed eager, to mate with members of other subspecies of *Felis silvestris*, even those which, like the Scottish wildcat, have become genetically distinct following tens of thousands of years of separation from the domestic cat's wild ancestors in the Middle East. For some reason, the offspring of such liaisons, while themselves perfectly capable of reproducing, are rarely incorporated into the pet population nowadays; rather, those that survive adopt the wildcat lifestyle. Presumably there is some kind of genetic incompatibility in these hybrids that suppresses the full expression of the genes that would enable them to become socialized to people. Evidently no such incompatibility existed between the earliest domestic cats and the wildcats around them.

As humans began to take cats on their travels, so those cats would have encountered local wildcats belonging to the *lybica* subspecies, and assimilated some of their genes. With no biological barrier to mating, tame females must have been successfully courted by wildcat males. Sometimes, as has been seen recently among cats in Scotland, the resulting kittens would have taken after their father and been unhandleable. Occasionally, however, the kittens would have been easy to tame and stayed with their mother, fusing with the domestic population. This cannot account for all the genetic diversity in today's cats, however, because this process accounts only for new genetic material being introduced from male wildcats. Domestic cats bear the hallmarks of descent from many wildcat males, certainly, but also from about five different individual wildcat females, each of which can be located with some certainty in either the Middle East or North Africa.[18] It is possible that each of these five individuals was domesticated separately, each by a different culture in a different location, and that their descendants were subsequently—perhaps hundreds or even thousands of years later—traded among cultures, until all the genomes became mixed together. However, such an explanation may give too much agency in this process to humans and not enough to the cats themselves.

The ability of those early cats to interbreed with their wild neighbors is what gave them that extra genetic diversity. Every now and then, a tame, semidomesticated male, lured by the scent and mating calls of a wild female, would have escaped and mated. A few of the resulting offspring could have carried the right genes to be easily tamable; some of these might have been found and adopted as pets by local women or children, and then raised to mate with other domestic cat males. This need not have happened very often: only four or five of these, in addition to the original founding female, have their descendants in today's pets.

The prehistory of the cat is thus the result of many fortuitous interactions among human intent, human affection for cute animals, and cat biology. It was a far more haphazard process than the domestications of other animals—sheep, goats, cattle, and pigs—that took place at the same time. Domestic dogs of different types were already emerging, showing that people of the time could channel their domestic animals into forms that were more useful and easier to handle than their predecessors. Yet for thousands of years, the cat remained an essentially wild animal, interbreeding with the local wild populations—such that in many places, the tame and the wild must have formed a continuum rather than the polar opposites they are today. Moreover, wild and domestic cats would have been almost identical in appearance, and distinguishable only by their behavior toward people. To earn the tolerance of their human hosts, cats had to be effective hunters: any cat that allowed mice to flourish in its owner's barn, or perhaps let a snake into the house to bite and poison one of the family, would not have lasted long. Docility, low reactivity, and a dependence on humans to take the lead—prized characteristics in other domesticated animals—would not have done the cat any favors.

Nevertheless, the first artistic and written records that we have of cats depict them firmly as part of the family, so they evidently inspired feelings of affection in humans at least toward the end of this predomestication phase. Only now is feline science allowing us to understand how and why this has come to be.

CHAPTER 2

The Cat Steps Out of the Wild

We will never pin down the precise time or place where cats gave up the wild for good. There was no single, dramatic domestication event, the brainstorm of some early miller who realized that cats were the ideal solution to his rodent problems. Instead, the cat gradually insinuated itself into our homes and hearts, changing from wild to domestic in fits and starts, over the course of several thousand years.

This progression probably saw many failed beginnings, as one person after another hand-reared some especially tractable kittens in different locations in the Middle East and northeast Africa. These people probably bred two or three litters of cats, and then either lost the habit or lost the cats themselves, which reverted to the wild. These false starts would have occurred from time to time over a period of perhaps five thousand years, beginning when humankind first started storing food for long enough to attract mice and other vermin, some eleven thousand years ago. Some might have lasted for just a few generations of cats, while others may have persisted for decades, perhaps even a century or two. However, temporary liaisons such as these leave little trace in the archaeological record, especially where wild and tame cats lived side by side and differed only in their behavior.

We have only one well-documented example from this period of a close relationship between humans and cats. In 2001, archaeologists from the Natural History Museum in Paris had been excavating a Neolithic village at Shillourokambos in Cyprus for more than a decade when they discovered a complete cat skeleton, dating to around 7,500 BCE, buried in a grave.[1] That the skeleton was still intact and that the grave had been dug deliberately both suggested that the burial had been far from accidental; moreover, the cat lay within fifteen inches of a human skeleton, whose grave also contained polished stone tools, flint axes, and ochre, indicating that this was a human of high status. The cat was not fully grown, probably less than a year old when it died, and although nothing else indicated that it had been killed deliberately, the cat's age suggests that this is indeed what happened.

Cat burial in Cyprus

We can only guess at the relationship between this cat and this person, but it resulted in their being buried near to each other. Unlike some dog burials of the time, the human and cat were not placed in physical contact, suggesting that the cat was not a treasured pet; instead, there was an arm's-length relationship between the two. Yet the very fact that this cat was buried with such deliberation suggests that someone, perhaps the person in the grave or a surviving relative, valued it highly.

This single cat skeleton gives us a glimpse into an early relationship between cats and humans, but also poses more questions than it answers. No burials of cats have been recorded from the mainland Middle East until thousands of years later. If cats were fully domesticated pets during this period, some of them should have been buried with the same formality as dogs routinely were at the time. The initial domestication of the cat may just possibly have taken place on Cyprus, and some might have been subsequently exported back to the Middle East to form one of the nuclei that eventually led to today's pets, but we have no evidence to support this idea. More likely, the Cyprus burial represents an anomaly: a very special human and his prized tame wildcat.

For the cat to make the leap into domestication, it almost certainly had to become an object of affection as well as utility: some of the ancestors of today's cats must have been pets in addition to pest controllers. We have little direct evidence for pet-keeping of any kind, apart from dogs, in the Neolithic cultures of the Eastern Mediterranean, but a number of present-day hunter-gatherer societies practice something like it, which may provide clues as to the process by which wildcats originally became first tame and then domestic. In both Borneo and Amazonia, women and children in such societies adopt newly weaned animals taken from the wild, and keep them as pets.[2] Since the habit of creating pets from young wild animals is found in societies that have never had contact with one another, we might consider this a universal human trait. If so, this could account for the possible adoption of wildcat kittens by people on the shores of the

Mediterranean, one of which may have been taken by its owner across the sea to Cyprus. The human skeleton buried alongside the cat is that of a man, so pet-keeping was possibly, and somewhat unusually, practiced by both men and women at that location and time.

If the first cats to live in human settlements were indeed tamed wildcats, they are unlikely the direct ancestors of today's domestic cats. In modern hunter-gatherer societies, young animals taken from the wild, whatever the species, are rarely kept for very long and do not usually breed in captivity. Rather, as they grow and their cuteness fades, they may be abandoned, driven away, or even eaten, if they are known to be tasty enough and local taboos allow it. For example, such a relationship exists today between the dingo and some Aboriginal tribes of Australia. The dingo is a not a true wild dog, but is descended from domestic dogs that escaped into northern Australia several thousand years ago and became successful predators, rather like the wildcats of Cyprus. Some Aborigines find dingo puppies irresistible and take them from the wild to keep as pets. However, as they grow into adolescence, these puppies become a considerable nuisance, stealing food and harassing children, so they are driven back into the wild. We can easily imagine the affectionate relationship between humans and wildcats beginning in a similar way.

The first clear indications that cats had transformed themselves into pets come from Egypt, just over 4,000 years ago. At that time, cats started to appear in paintings and carvings. It is not always clear what species these cats are—some, especially those without tabby markings, could easily be jungle cats. Indeed, we have evidence that the Egyptians had already been keeping tame jungle cats for hundreds of years—a skeleton of a young jungle cat recovered from a 5,700-year-old grave had healed fractures in its legs, indicating that it must have been nursed for many weeks before it died.[3] There is no indication that these jungle cats were any different from their wild counterparts, so they had not been domesticated in the sense that their genetic makeup had been altered by their association with humankind. Other cats, while clearly striped and therefore presumably

Tethered cat—Egypt 1450 BCE

Felis silvestris, appear in outdoor scenes, often reedbeds, alongside other local wild predators such as genets and mongooses, making it more likely that they are wildcats than domestic. Even the cats that appear in indoor scenes are sometimes depicted wearing collars, and so could be tamed wildcats rather than domestic cats. However, early in the Middle Kingdom, about 4,000 years ago, a set of hieroglyphs—"miw," in translation—was created specifically for the domestic cat. Not long after this, Miw was adopted as a name for girls, a further indication that by that time the domestic cat had become an integral part of Egyptian society.[4]

We see hints of pet cats in Egypt going back a full 2,000 years earlier than this, into the Predynastic era. The tomb of a craftsman, constructed some 6,500 years ago in a town in Middle Egypt, contained the bones of a gazelle and a cat. The gazelle was probably placed there to provide the craftsman with food for the afterlife, but the burial of the cat, perhaps his pet, is reminiscent of the similar burial on Cyprus some 3,000 years earlier. In a cemetery at Abydos in Upper Egypt, 500 miles south of the Mediterranean, a 4,000-year-old tomb contained the skeletons of no fewer than seventeen cats. Alongside them were a number of small pots that had probably contained milk. Although the reason for the burial of so many cats in one place is obscure, that they were buried with their food bowls indicates that these cats must have been pets.

These early pets may have come from stock domesticated locally, or they may have been imported from elsewhere. If cats were indeed domesticated farther north in the Fertile Crescent, or even possibly in Cyprus, long before the rise of Egypt as a center of civilization, they were likely traded around the region, possibly as exotic novelties. This theory accounts for the scarcity of evidence for domestic cats in Predynastic Egypt. The cats that found their way there would have been valued possessions because their owners had paid well for them, but they might have been too few in number to sustain themselves as domestic animals. Most would have been unable to locate a domesticated member of the opposite sex, and would have mated with, or been mated by, a local wildcat or possibly a tamed wildcat. In this way, the genetic differences between domestic and wild cats at that time would have been swiftly diluted by the wild versions, and each subsequent generation would have become less and less likely to accept a domestic lifestyle.

The domestic role of the cat in Egypt becomes much clearer over the next 500 years, probably reflecting the emergence of a local, self-sustaining, domestic population. Cats sitting in baskets—surely a sign of domesticity—appear in Egyptian temple art between about 4,000 and 3,500 years ago. In paintings dating from about 3,300 years ago, the cats are often depicted sitting—unrestrained—beneath the chair of an important member of the household, often the wife. (The animal beneath the husband's chair is commonly, and perhaps not unexpectedly, a dog.) In one painting from about 3,250 years ago, not only do we see

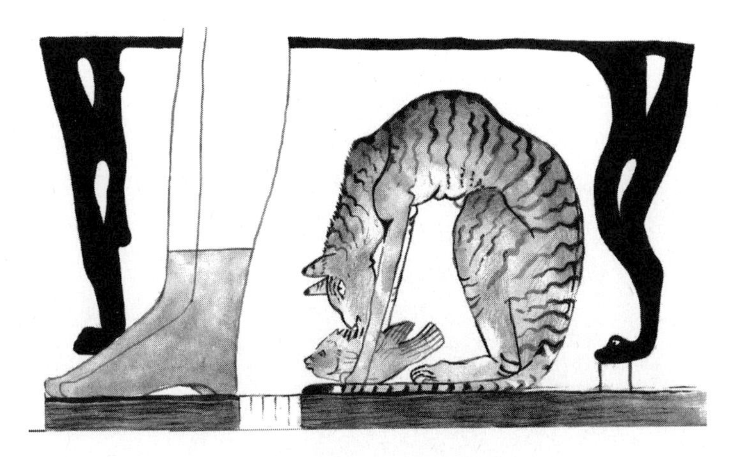

Pet cat—Egypt 1250 BCE

an adult cat sitting under the wife's chair, but also her husband has a kitten on his lap. Members of the Egyptian nobility were evidently deeply attached to their cats, including the eldest son of Pharaoh Amenhotep III, who died at the age of thirty-eight during the same period. He was so fond of his cat Osiris, Ta-Miaut (which translates as Osiris, the she-cat), that when she died he not only had her embalmed, but also had a sarcophagus (stone coffin) carved for her.[5]

Almost all of these cats are depicted in aristocratic surroundings, supporting the idea that cats were still exotic pets, reserved for the privileged few. We find little direct evidence of cats in the homes of working people at this time: this, however, is largely because the tombs and temples, many sited on the edge of the desert, are so much better preserved than ordinary people's dwellings, which were nearer to the Nile. Luckily, the artists who worked on creating the tombs and temples between 3,500 and 3,000 years ago left behind drawings, presumably done for their own pleasure; many are humorous and cartoonlike, in contrast to the formal drawings required for temple decoration. Many of these drawings depict cats—some in ordinary domestic situations, and others in more imaginary contexts, such as an image of a cat carrying a pack on a stick over its shoulder, strangely reminiscent of the later English

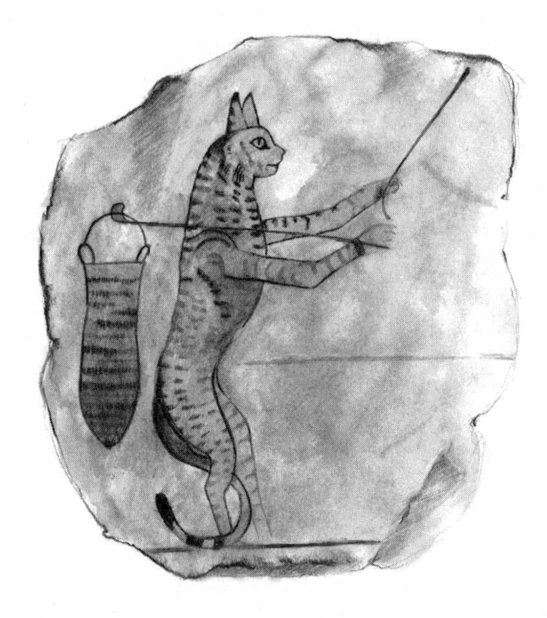

Cartoon cat on a limestone tablet—Egypt 1100 BCE

folktale of Dick Whittington's Cat. These drawings help to confirm that by this time, pet cats were widespread and commonplace in Egypt.

We have good evidence that the Egyptians, as well as treasuring their cats as companions, also regarded them as useful. Some depictions of cats from about 3,300 years ago show them apparently accompanying Egyptians on hunting trips, but these depictions are almost certainly fanciful; we have no evidence of any other culture using cats for this purpose—and just imagine trying to do anything like this with one of today's domestic cats! It is far more likely that cats were becoming domesticated for their ability to keep pests, such as imported house mice and indigenous wild rodents, out of the granaries and other food stores on which the Egyptian economy depended. One such pest was the Nile rat, smaller and chubbier than the more familiar brown rat, but no less devastating. Agriculture in the Nile valley depended on the annual flooding of the arable land either side of the river, refreshing the soil with much-needed nutrients washed downstream. This flooding would also have driven Nile rats, searching for food and shelter, from their communal burrows up to higher ground, where granaries were sited.[6] Cats would have been useful deterrents against such invasions.

The Egyptians seem to have valued cats not only for their ability to keep rodent pests away, but also for their expertise in killing snakes. Venomous snakes were a source of considerable anxiety in ancient Egypt: the Brooklyn Papyrus, dating from about 3,700 years ago, is concerned largely with remedies for snakebites and the venoms of scorpions and tarantulas. The Egyptians used both the mongoose and the genet as snake-exterminators, but these were individually tamed from the wild; the cat was the only domesticated animal capable of killing snakes.[7] The historian Diodorus Siculus, recording life in Egypt more than a millennium later, wrote, "The cat is very serviceable against the venomous stings of serpents, and the deadly bite of the asp."[8]

Clearly the Egyptians regarded domestic cats as useful protection against poisonous snakes, although we do not know the extent to which this was based on their actual effectiveness in preventing snakebites. Today's cat owner might be surprised to learn that Egyptian cats would have attacked snakes rather than run away from them. Pet cats rarely kill snakes in Europe—the only reptiles recorded to be eaten there are lizards—and in the United States, cats are known to kill and eat lizards and non-venomous snakes. Only Australia has records of cats killing venomous snakes; many feral cats in Australia kill and eat more reptiles than they do mammals. We have few studies of cats' diets from Africa, and none from Egypt, but English scholars working in Egypt in the 1930s reported seeing cats killing horned vipers, and menacing, if not actually killing, cobras.[9] It is highly unlikely that cats were ever specially bred to prey on snakes—mongooses are much more skilled at this—but incidents like these may have made a lasting impression on the ancient Egyptians who witnessed them.[10] The Egyptians must have used cats mainly to kill mice and other rodents, both in the home and in granaries—a function presumably too mundane to feature in Egyptian art or mythology.

In the next stage of the domestic cat's evolution as a controller of vermin, it encountered a new enemy: the black rat, *Rattus rattus*. Originating from India and Southeast Asia, this pest had spread east along trade routes to the civilizations of Pakistan, the Middle East, and Egypt by about 2,300 years ago. From there, it hitched a ride on Roman trading vessels, reaching Western Europe by the first century AD. Black rats are more generalist feeders than house mice, eating all kinds of stored foods as well as feeds prepared for domestic livestock. Additionally, they are carriers of disease, and were recognized as such by both the Greeks and the Romans. Had cats been unable to control this new threat, humans might well have given them the cold shoulder. However, the cats of 2,000 years ago, larger than today's cats, appear to have risen to the challenge. A peculiar cat burial on the Red Sea coast 1,800 years ago shows that at least some cats of that era were effective predators of rats. The cat in question was a large young male, typical of the cats of the time but something of a giant by today's standards. Before burial, it had been wrapped in pieces of a woolen cloth decorated in green and purple, beneath a linen shroud similar to that of an Egyptian mummy. However, the cat was not mummified in the conventional way, for its intestines were not removed. Examiners found in its stomach contents the bones of at least five black rats, and at least one more, farther down the gut.[11] It is unclear why this cat died and why it was afforded such an elaborate burial, but it may have been special to its owner because of its champion ratting skills.

Egypt esteemed cats for their roles as pets and pest controllers, and also endowed them with spiritual significance: starting about 3,500 years ago, cats became increasingly prominent in Egyptian cults and religion. Depictions of cats begin to feature on tomb walls; depictions of the sun god occasionally have the head of a cat instead of a human, and are referred to as "Miuty." The lioness deities Pakhet and Sekhmet (the latter also associated with the caracal) and the leopard goddess Mafdet, although clearly based on big cats known to the

Egyptians, nevertheless gradually became associated with domestic cats, presumably because they would have been to most people the most familiar and accessible members of the cat family.

Bastet was the goddess with which the ancient Egyptians came to associate the domestic cat most closely. Worship of Bastet originated in the city of Bubastis, in the Nile delta, some 4,800 years ago. She originally took the form of a woman with a lion's head, carrying a serpent on her forehead. Some 2,000 years later, the Egyptians began associate her with smaller cats; presumably this followed with the arrival of domestic cats in the city, or even a new local domestication. During this period, Bastet still had the head of a lioness, but was sometimes depicted with several smaller, presumably domestic cats as her attendants. Within the next 300 years, some 2,600 years ago, her lion-goddess identity apparently mutated to resemble the domestic cat more closely. Originally a simple goddess who protected humankind against misfortune, she later became associated with playfulness, fertility, motherhood, and female sexuality—all characteristics of domestic cats. Her popularity spread to other parts of Egypt, especially during the Late Period and the Ptolemaic Era (2,600 to 2,050 years ago), as the Egyptian empire gradually crumbled. Her annual feast day was for a time the most important in the calendar, as witnessed by the Greek historian Herodotus:

> Now, when they are coming to the city of Bubastis they do as follows: they sail men and women together, and a great multitude of each sex in every boat; and some of the women have rattles and rattle with them, while some of the men play the flute during the whole time of the voyage, and the rest, both women and men, sing and clap their hands; and when as they sail they come opposite to any city on the way they bring the boat to land, and some of the women continue to do as I have said, others cry aloud and jeer at the women in that city, some dance, and some stand up and pull up their garments. This they do by every city along the riverbank; and when they come to Bubastis they hold festival celebrating great sacrifices, and more wine of grapes is consumed upon that festival than during the whole of the rest of the year.[12]

Presumably because of their association with this cult, the Egyptians seem to have been extremely protective of cats, in ways that may seem absurd to us today. Herodotus reported that when a pet cat died from natural causes, all members of the household shaved their eyebrows as a mark of respect. He even reported seeing Egyptians struggling to prevent cats from entering a burning building in preference to putting out the fire itself.[13] This veneration of cats clearly persisted over time. Some 500 years later, when Egypt was part of the Roman Empire, Diodorus Siculus wrote:

> If any kill a cat, whether willfully or otherwise, he is certainly dragged away to death by the multitude. For fear of this, if any by chance find any of these creatures dead, they stand aloof, and with lamentable cries and protestations, tell everybody that they found it dead. . . . It so happened that upon a cat being killed by a Roman, the people in tumult ran to his lodging, and neither the princes sent by the king to dissuade them, nor the fear of the Romans, could deliver the person from the rage of the people, though he did it [presumably, the killing of the cat] against his will.[14]

In contrast to this behavior, the Egyptians routinely practiced infanticide toward their cats. Herodotus wrote, "They either take away by force or remove secretly the young from the females and kill them (but after killing they do not eat them)."[15] This account of a convenient method of population control suggests that by his time, and probably much earlier, domestic cats were breeding freely as a self-contained population, more or less isolated from their wild counterparts, and that far more kittens were being born than were needed to become either pest controllers or pets. To modern sensibilities, this brutally pragmatic culling of kittens may seem callous, but before the advent of modern veterinary medicine, it was the simplest way to keep cats' numbers within reasonable bounds. It is presumably least distressing for the perpetrator to dispose of kittens before their eyes open and their faces take on their characteristic appeal. In societies where cats are useful pest controllers first and pets second, this has remained standard practice into the modern era. Describing attitudes toward cats in 1940s rural New Hampshire, Elizabeth Marshall Thomas noted that:

> Farm cats, after all, are neither pets nor livestock. . . . When the cat population got too high for a farmer's liking, the cats were simply put into bags and gassed or drowned. To care for a group of animals for a time, and then to suddenly round them up and dispatch them without warning, is after all what farming is all about.[16]

Even in the twenty-first century, acceptable behavior toward cats varies widely. Some people view cats as individuals with rights, but others see them as tools that can be discarded when they are no longer useful.

The ancient Egyptians, with their deep reverence for cats, added an additional dimension to cat culture that is abhorrent to us today: the cat as sacrificial object. Cats not only formed an important part of the Egyptian pantheon, but were also ritually buried in large numbers—almost certainly millions. The Egyptians, who placed great emphasis on the afterlife, developed the process of mummification some 4,000 years ago as a way to preserve corpses, both human and animal.

Initially, mummification of cats seems to have been reserved for treasured pets. A mummified cat is depicted on the sarcophagus of Ta-Miaut, so presumably this cat had been mummified and the sarcophagus constructed specifically to house the mummy. This practice probably continued for many hundreds of years, but the numbers of cats involved were tiny compared with the millions that were mummified as offerings to various cat deities.

The production of "sacred animals" became a major industry in Egypt between 2,400 and 2,000 years ago. Small cats were by no means the only animal involved: mummies also included lions and jungle cats, cattle, crocodiles, rams, dogs, baboons, mongooses, birds, and snakes. Sometimes staggering numbers of animals were treated this way: for example, more than 4 million mummified ibis, a medium-sized wading bird that the Egyptians bred in captivity, were recovered from catacombs at Tuna el-Gebel, and an additional 1.5 million from Saqqara.

Modern analyses show that cat mummification was often performed to a high standard, including many of the techniques used for the mummification of human corpses. To preserve the body, the intestines were removed and replaced with dry sand.[17] Once the corpse was prepared, the cat was wrapped in layers of linen bandages. These were often treated with a preservative, such as natron, a natural desiccant and preservative that forms on the beds of dried-out tropical lakes. Other times, the embalmers used mixtures consisting of animal fats, balsam, beeswax, resins from trees such as cedar and pistacia, and occasionally bitumen brought from the Red Sea coast, more than a hundred miles from where the cats were kept.[18]

Cat "mummies" varied considerably in external appearance, presumably reflecting the tastes and finances of the prospective purchaser. Some were a simple bundle, with perhaps a simple string of glazed pottery beads for decoration, but others had an outer layer of linen applied with decorative patterns. A "head" might have been molded around and above the actual skull, using clay and plaster-soaked linen, or a bronze head might have been added; some were crude, but others depicted every whisker. Many were placed in simple rectangular wooden coffins, but for others, cat-shaped wooden caskets were constructed, decorated with plaster, and then painted, and sometimes even gilded. Inlaid beads were used to represent the cat's eyes; overall, they must have been astonishingly lifelike when first constructed.

What is more, the sacrificial cats were bred specifically for this purpose. Remains of catteries have been found adjacent to the temples of all the deities associated with cats or other felids. There is little doubt that these cats were deliberately killed for mummification, since X-rays of the mummies show that their necks had been dislocated, and others were probably strangled.[19] Some were killed when they were still kittens, at two to four months old, while others were fully grown, at nine to twelve months: presumably the purveyors of such a commercial operation saw no benefit in feeding a cat for any

Mummified cats and a casket

longer than this unless it had been earmarked for breeding. The mummies would be sold to visitors to the temple, who would then leave them there as offerings to the appropriate deity. As sufficient numbers accumulated, the priests would remove them in batches to dedicated catacombs, where many remained, well preserved, until the pillaging of the cemeteries in the nineteenth and twentieth centuries.

We will never know how many cats were sacrificed this way. The archaeologists who discovered these sites wrote of vast heaps of white cat bones, and dust from disintegrating plaster and linen blowing across the desert. Several other cemeteries were excavated wholesale, and their contents ground up and used as fertilizer—some was used locally, some was exported. One shipment of cat mummies alone, sent to London, weighed nineteen tons, out of which just one cat was removed and presented to the British Museum before the remainder were ground into powder. Out of the millions that were mummified, only a few hundred now survive in museums, and these come from a mere handful of the many cemeteries constructed over a period of several hundred years. As such, these mummies may not be entirely representative of the cats of ancient Egypt.

Examination of some of the few remaining mummies using forensic techniques has revealed much about the animals preserved inside, providing insight into the relationships that the Egyptians had with their cats. All the cats were "mackerel"-striped tabbies, the same as the wild *lybica*; none were black, or tabby and white, and none had the blotched tabby pattern more common than the striped version in many parts of the world today.

Evidence for such color and pattern variations did not appear in domestic cats until later, and not in Egypt. Each of these changes in appearance is caused by a single mutation that is also common in wild felids. For example, the so-called king cheetah, once thought to be a distinct species, simply has a blotched tabby coat rather than the normal spotted variety. Black ("melanistic") forms of felids abound: they have been recorded from lions, tigers, jaguars, caracals, pumas, bobcats, ocelots, margays, and servals. In the wild, their black color is a handicap because it destroys the camouflaging effect of their normal coat, so they leave few offspring and the gene responsible disappears from the population.[20]

Given this, it seems odd that despite a possible 2,000-year history of domestication, none of these color varieties are apparent in the Egyptian mummified cats; they all seem to have become established within the next two millennia. Perhaps the Egyptians actively discouraged these "unnatural" cats on the rare occasions when the mutations occurred, possibly for reasons connected with religion.

Some of the cats in ancient Egypt may have been ginger or ginger-tabby ("torbie") mixes (see box on page 41, "Why Ginger Cats Are Usually Boys"). Some of the wall paintings are a more orange shade of brown than the normal grayish-brown of *lybica*, although this may be the consequence of some artistic license, or due to yellowing of the pigments over the centuries. Ginger cats are more common in the Egyptian port of Alexandria and the Egyptian-founded city of Khartoum

Why Ginger Cats Are Usually Boys

The mutation that causes a cat's coat to be ginger rather than the usual shades of brown and black is inherited differently from other coat colors. In mammals, most genes obey the "dominance" rule: to affect the appearance of the animal, the "recessive" version has to be present on both chromosomes, one inherited from the mother and one from the father; otherwise, the other "dominates." Usually, animals with one dominant and one recessive gene are indistinguishable on the outside from animals with two dominant versions. However, there is one major exception: if a cat carries one orange and one brown version of the gene, then *both* appear in the coat, in random patches: in one part of the skin, the chromosome with the orange version has been switched on, and in another it is the brown-black pattern that "wins," producing a tortoiseshell-tabby (or "torbie") cat. The precise color of the patches depends on other coat-color genes. If the cat also carries (two copies of) the black mutation, then the brown patches are black, so their tabby pattern is obscured (like a regular black cat), while the orange patches, on which the black mutation has no effect, are orange and yellow with the tabby pattern still visible, producing a tortoiseshell or calico.

Second, the gene is carried on the X chromosome. Female cats have two X chromosomes, and males only one, paired with the much smaller Y chromosome that makes them male but carries no information about coat color. Thus, for a female cat to be orange, it must carry the orange mutation on both chromosomes. If it has only one, it will have a tortoiseshell coat. Although tortoiseshells are much more common, ginger cats can be female. Almost without exception, males are either orange, or they're not; in fact, tortoiseshell males do crop up from time to time—they have two X chromosomes *and* a Y chromosome, the result of an abnormal cell division. A common misperception holds that ginger, or "marmalade," cats are always male—hence the phrase "ginger tom."

than anywhere else in northeast Africa or the Middle East, suggesting that ancient Egypt was indeed where the orange mutation originally became incorporated into the domestic cat population, before spreading from there to the rest of the world.[21] Although orange cats look more conspicuous than tabby cats, and may seem less effectively camouflaged, today's orange cats can be very successful hunters, especially in rural areas. Once the mutation had occurred, there seems no reason for it not to have spread through the cat population.[22]

We also know that the mummified cats were about 15 percent larger than modern pet cats.[23] In almost every other domestic species—cattle, pigs, horses, and even dogs—the early domesticated forms are significantly *smaller* than their wild counterparts, mainly because smaller individuals are easier to handle. However, this principle may not apply to the cat, which was small relative to man to begin with. More surprising, the mummified cats were also 10 percent larger than African wildcats are today. It may be that the Egyptians deliberately favored large wildcats because they were more effective rodent controllers, and that domestic cats have subsequently become smaller as they gradually transformed from full-time pest controllers into pets.

The attitudes of the ancient Egyptians toward their cats seem paradoxical, almost unthinkable, to modern sensibilities. To the Egyptians, some cats were revered pets, many more were simply pest controllers, used by rich and poor alike, but, uniquely, many were bred specifically to be killed as sacrifices. Apart from the last, all this is not very different from the way that cats were regarded in Europe and the United States in the first half of the twentieth century. Indeed, the Egyptian habit of having elaborate coffins made for favorite pet cats mirrors today's cat cemeteries.

Undoubtedly, the Egyptian association between cats and religion seems the most foreign to us. The worshipers who bought ready-prepared mummies as offerings at the temple can hardly have been unaware of what the contents of those mummies were, for the breeding premises and production line for mummies were both nearby, and would have been evident from their smell alone. Presumably these sacrificial cats were regarded in some way as "different" than household cats, even though they would have been genetically indistinguishable. Perhaps this was reinforced by their being bred in purpose-built catteries. Cats would have been just as prolific breeders in those days as they are now, and mummy manufacturers would have found trapping young feral cats easy. Since both law and custom forbade this, separate breeding of cats must have been the only solution. It is possible that access to the premises used was prohibited to all but the priests who looked after the "sacred" cats, which were never seen by the worshipers until the cats had been mummified, thus maintaining a distinction between household and sacred cats, even though they were otherwise identical.

The mummies themselves show a paradoxical concern for welfare during life, but no trace of concern for life itself; the procurers of these mummies evidently took great care of their cats, but then killed them in huge numbers. Judging from the sizes of the cats, they were evidently well nourished. Finding sufficient high-quality meat and fish for such large numbers of animals could not have been easy. Although it is not entirely clear how all the cats were killed, it seems most likely that they were strangled in some kind of prescribed, ritual manner. Moreover, although the production of mummies must have been a lucrative business, there seem to have been few attempts to cheat the worshipers; almost all the mummies that were produced to look like cats actually do contain a complete cat skeleton, even though it would presumably have been more profitable to wrap a bundle of reeds in linen and pass it off as a cat mummy. The entire process seems to have been carried out according to strict rules. These rules protected not only the worshipers from the purchase of fake mummies, but also the cats, which were well-fed and cared for, at least by the standards of the time, up to the point when they were sacrificed.

The cats of ancient Egypt were likely the main ancestors of modern-day cats, as several of their qualities attest. We have no credible evidence for large-scale domestication of the cat anywhere else in the world before the birth of Christ. These domestic cats had the wildcat's striped tabby coat, so would have been distinguishable from genuine wildcats only by their affection for, rather than fear of, people. Some were pets, certainly in well-to-do households and most probably in many others. Most would have been useful in keeping food stores and granaries reasonably free of rodent pests. Household cats were venerated, at least during the last few hundred years of Egyptian civilization, as indicated by the illegality of killing them, and the rituals performed when one died.

From about 2,500 years ago, keeping cats gradually became more widespread around the eastern and northern shores of the Mediterranean, as Egypt came first under Greek and then Roman influence. Historians have traditionally ascribed the slow northwards spread of the cat to laws that prohibited the export of cats from Egypt. Some accounts even tell of the Egyptians sending out soldiers to retrieve and repatriate cats that had been taken abroad.[24] However, these laws were almost certainly symbolic, connected to cat worship. The cat's independence, hunting ability, and rapid rate of reproduction would have made it impossible for Egyptian authorities to prevent domesticated cats from spreading along their trade routes.

As cats spread out of Egypt, they must have come across, and interbred with, wild and semi-domesticated cats in other areas of the Eastern Mediterranean. As the Cyprus cats attest, there must have been tamed cats in other parts of the Middle East for thousands of years before the Egyptians began to transform them into domesticated animals. Egyptian paintings from about 3,500 years ago show cats onboard ships, and these cats could plausibly have been either immigrants or emigrants. Between 3,200 and 2,800 years ago, trade in the eastern Mediterranean was dominated by the seafaring Phoenicians (who may even have domesticated their own wildcats as pest controllers), operating out of several city-states in what is now Lebanon and Syria. The Phoenicians probably introduced *lybica* cats, either tamed or partly domesticated, to many of the Mediterranean islands and to mainland Italy and Spain. The spread of the cat was probably delayed not by Egyptian laws, but more by the presence in Greece and

Rome of rival rodent controllers: tamed weasels and polecats (the latter becoming domesticated as the ferret).

The domestic cat's migration north from Egypt toward Greece is not well documented. In the Akkadian language, spoken in the eastern part of the Fertile Crescent, separate words for domestic cat and wildcat appear about 2,900 years ago, so domestic cats had probably spread into what is now Iraq by this period.

Domestic cats were likely common in Greece, at least among the aristocracy, some time before this. We know this from coins minted for use in two Greek colonies. They were made about 2,400 years ago, one for Reggio di Calabria, on the "toe" of Italy opposite Sicily, and the other for Taranto on the "heel," and both depict their founders, some 300 years previously. Although these were different people, the coins are remarkably similar and may both refer to the same legend. Both show a man sitting on a chair, dangling a toy in front of a cat, which is reaching up with its forepaws. That this man is shown with a cat rather than the more usual horse or dog suggests that pet cats were initially unusual in Greece, possibly exotic imports from Egypt, and their possession an indicator of status. A bas-relief carved in

Greek coin—Italy 400 BCE

Athens at about the same time shows a cat and a dog about to fight, but the cat is leashed, implying a tamed rather than domestic animal.

Domestic cats probably became common in Greece and Italy about 2,400 years ago. Some of the earliest clear evidence comes from Greek paintings, in which cats are shown unleashed and relaxed in the presence of people. Cats also began to be depicted on gravestones, presumably as the pets of the people buried there. Furthermore, by this time, the Greeks had a word specifically for the domestic cat—"aielouros," or "waving tail." In Rome, paintings began to appear showing cats in domestic situations—under benches at a banquet, on a boy's shoulder, playing with a ball of string dangled from a woman's hand. As in Egypt, cats in Rome were women's pets, men generally preferring dogs. And as "Miw" had been adopted as a name for girls in Egypt, so "Felicula"—little kitten—became a common name for girls in Rome by about 2,000 years ago. In other parts of the Roman empire, "Catta" or "Cattula" were used, the former originating in Roman-occupied North Africa.

As in Egypt, once the cat had been domesticated, it began to be associated with goddesses—particularly Artemis in Greece and Diana in Rome. The Roman poet Ovid wrote of a mythical war between gods and giants in which Diana escaped to Egypt and transformed herself into a cat to avoid detection. Thus, cats became widely associated with paganism, a link that would eventually lead to their persecution in the Middle Ages.

As sea routes opened up between the Middle East, the Indian subcontinent, and the Malay peninsula and Indonesia, so cats—for the first time—were transported out of the native areas of their wild ancestors. Roman traders were probably responsible for carrying cats to India, by sea, and later to China through Mongolia, along the Silk Road. Cats were established in China by the fifth century AD, and in Japan about a hundred years later.[25] In both countries, cats became valued especially for their ability to protect the valuable silk moth cocoons from attack by rodents.

The characteristic Southeast Asian type of domestic cat—lean-bodied, agile, and vocal—is not, as was once thought, a separate domestication of the Indian desert cat, *Felis ornata*. Although archaeological evidence is sparse, the DNA of today's street cats across the Far East—whether from Singapore, Vietnam, China, or Korea—shows that they have the same ultimate ancestor as European cats: *Felis lybica* from northeast Africa or the Middle East. So do all the "foreign" (Far Eastern) purebreds, such as the Siamese, Korat, and Burmese.[26] No barrier prevents domestic cats from breeding with Indian desert cats, but evidently their offspring rarely make suitable pets: although wildcats in central Asia carry domestic DNA, their domestic counterparts show no trace of wild DNA.

Precisely dating the appearance of domestic cats in Southeast Asia is impossible, and each population seems to have developed in isolation from the others. The DNA of street cats in Korea is fairly similar to that of their counterparts in China, and to a lesser extent, those in Singapore, but the cats of Vietnam are considerably different, implying that there has been little transfer of cats between these countries since they first arrived. The street cats of Sri Lanka are different again, more closely resembling those of Kenya than anywhere in Asia, perhaps due to transfer of ship's cats across the Indian Ocean.

History paints a conventional picture of the origins and spread of the domestic cat up to the time of the birth of Christ, but this picture is at odds with what we can deduce from biology. Conventional accounts have emphasized human intervention, and presume that the domestication of the cat was a deliberate process. From the cat's perspective, a different picture emerges: one of a gradual shift from wild hunter to opportunist predator, and then, via domestication, to parallel roles as pest controller, companion, and symbolic animal.

Biologists would observe that at each stage, cats were simply evolving to take advantage of new opportunities provided by human activities. Unlike the dog, which was domesticated much earlier, there would have been no niche for the cat in a hunter-gatherer society. It was not until the first grain stores appeared, resulting in localized concentrations of wild rodents, that it would have been worth any cat's while to visit human habitations—and even then, those that did must have run the risk of being killed for their pelts. It was probably not until after the house mouse had evolved to exploit the new resource provided by human food stores that cats began to appear regularly in settlements, tolerated because they were obviously killing rodents and thereby protecting granaries.

As the practice of agriculture spread, so did the cat, encountering new challenges as it met new pests—for example, the Nile rat in Egypt and, later, the black rat in Europe and Asia. The cat had rivals for the role of vermin exterminator: other carnivores of similar size were tamed, including various members of the weasel family, and the genet and its cousin the Egyptian mongoose. Of these, the ferret was eventually domesticated from the weasels, and the mongoose, most effective as a controller of snakes, was introduced to the Iberian peninsula for that purpose by the occupying Caliphates as late as 750 AD.[27] These various rivals existed in different combinations in different places for many centuries, and it is not clear why the domestic cat was the eventual winner, with the ferret as the only runner-up. Very probably, cats did not have the edge as vermin controllers. The answer, therefore, must lie elsewhere, possibly in the cat's biology. The connection between cats and religion is unlikely to have been crucial, since Egyptians sometimes venerated both mongoose and genet as well.

More likely, the cat managed to become more profoundly domesticated than any of its rivals. However, which was the cause and which the effect? Are cats more "trustworthy" and predictable than ferrets because they have evolved ways of communicating with humans, or is it the other way around? Since we do not know precisely how the domestic cat's direct ancestors behaved, such questions are impossible to answer. Nevertheless, the cat's capacity to evolve not only into a pest controller but also into a pet animal—its present-day roles—must have been central to its success in the first 2,000 years of its domestication. So what set the cat apart on its millennia-long journey into our homes?

Here, the cat's involvement in Egyptian religion may well have been crucial. It is possible that the Egyptians' veneration of cats that gave the cat the time required to evolve fully from wild hunter to domestic pet; otherwise, it might have remained a satellite of human society and not an intrinsic part. It is even possible that the factories that produced the cat mummies forced the evolution of cats that could tolerate being kept in confined spaces and in close proximity with other cats, both qualities that are signally absent from today's strongly territorial wildcats but that are essential to life as an urban pet. Although of course most of the cats that carried the relevant genes died young—that was how they were being bred, after all—some must have escaped into the general population, where their descendants would have inherited an improved ability to deal with the close confines of urban society. Such changes take only a few decades in captive carnivores, as exemplified by the Russian fur-fox experiment that turned wild animals docile in just a few generations.[28] Is it possible that today's apartment-dwelling cat owes its very adaptability to the inhabitants of those gruesome Egyptian catteries?

CHAPTER 3

One Step Back, Two Steps Forward

The cats of Egypt two thousand years ago probably differed little in their behavior from modern cats. These cats were not yet so varied in appearance, and still consisted of a single, rather homogeneous population, with no pedigrees or distinctive types. Few obstacles would then have been apparent to stand in the path of the cat's inexorable rise to universal companion animal. However, this was not to be achieved for another 2,000 years, partly because the cat had only a single practical role. The dog, the cat's primary rival for human affection and attention, has adapted to serve many more functions—guarding, hunting, herding, to name but three—than the less malleable cat. Two other major factors also delayed the cat's rise to prominence, especially in Europe. First, the cat derived from a specialist carnivore, leaving it ill suited to scavenge for a living when prey was scarce. Second, the cat continued a long association with Egyptian religion—at first a blessing, granting the cat time to evolve into a domestic animal, but later a curse.

Surprisingly, Europeans' view of cats continued to be heavily influenced by Egyptian cat worship until as recently as about 400 years ago. Worship of Bastet (Bubastis) and other "pagan" deities associated with cats, such as Diana and Isis, was popular, in southern Europe especially, from the second to the sixth centuries CE. In some places, this worship lasted much longer: for example, Ypres, a Belgian city that celebrates its connections with cats to this day, outlawed cat worship only in 962 CE, while a cult based around the goddess Diana lingered in parts of Italy until the sixteenth century. Women managed most of these cults, focusing on motherhood, the family, and marriage. When Christianity began to establish itself as the preeminent religion of Europe, cats began to suffer from their affiliation with pagan practices.

The spread of cats from the eastern Mediterranean into Western Europe—and into all strata of society—was hastened by the custom of keeping cats on ships of all sizes. This habit likely arose for the practical purpose of keeping mice away from cargo, and soon became superstition, with many sailors refusing to sail in ships that had lost their cats. Ships often displayed carved images of cats on the prow to bring good luck.

Their obvious utility aside, cats must have continued to benefit from increasingly warm relationships with their human owners. In many ways, the vicissitudes of the cat's popularity over the past two millennia can be attributed to changes in the balance between two key influences: superstition and affection.

The Romans are often credited with introducing the domestic cat to Britain, but some evidence suggests that they had already arrived several centuries earlier. Cat and house mouse bones were found at two Iron-Age hill forts dating to 2,300 years ago, a few dozen miles apart in southern England. The cats were mostly young animals, with the youngest five newborn kittens. These hill forts, inhabited by as many as 300 people led by a local chieftain, were the focal point for several surrounding farmsteads, each of which supplied grain to large stores kept in the fort, perhaps holding twenty times as much as each individual farm. These stores inevitably became infested with mice, so cats would have been a useful addition to the domestic fauna.

These cats must have been brought from the Mediterranean, since it is highly unlikely that the local wildcats, though of the same species (*Felis silvestris silvestris*) could have been tamed, or even tolerated being in close proximity to people for long enough to produce kittens. Wildcats of European origin are notoriously wary of humans nowadays, and have probably always been so, judging from early Greek accounts of attempts to tame them. The most likely way domestic cats would have reached Britain is on Phoenician ships. The Phoenicians did not colonize Britain, but they did visit, chiefly to buy tin for the manufacture of bronze; and since they often carried cats on their vessels, the presence of domestic cats near the south coast of Britain is unsurprising. Indeed, the grain the Phoenicians brought to Britain on earlier voyages had quite possibly brought the house mice that the cats were called on to eliminate; indeed, the Phoenicians may subsequently also have supplied the cats as a way of clearing up the problem they had themselves created!

Cats likely became widespread in Britain during the Roman occupation. Archaeologists at the Roman town of Silchester, in what is now Hampshire, found a clay floor tile dating from the first century AD bearing the impression of a cat's foot. Presumably, the cat strayed into the drying yard before the tile had set. Other tiles unearthed at the same site bore the footmarks of a dog, a deer, a calf, a lamb, an infant, and a man with a hobnailed sandal—indicating that the Roman's reputation for quality control was not infallible.

The wane of Roman influence in northern Europe seems to have had little effect on the popularity of the cat: for the 500 years of the European Dark Ages, cats were highly valued for their rat- and mouse-catching skills.[1] Several statutes mention cats specifically, showing just how valuable they were. One such statute, from Wales in the tenth century, stated, "The price of a cat is fourpence. Her qualities are to see, to hear, to kill mice, to have her claws whole, and to nurse and not devour her kittens. If she be deficient in any one of these qualities, one third of her price must be returned." Note that this refers specifically to a female cat; toms were perhaps not regarded as equally precious. Fourpence was also the value of a full-grown sheep, goat, or untrained housedog. A newborn kitten was valued at one penny, the same as a piglet or a lamb, and a young cat at two. In a divorce, the husband had the right to take one cat from the household, but all the rest belonged to the wife. In Saxony, Germany, the penalty for killing a cat at this time was sixty bushels of grain (almost 500 gallons, or more than 1,500 kilograms), emphasizing the cat's value at keeping granaries free of mice.

Cats must also have helped to slow the spread of the rat-borne bubonic plague that swept through Europe in the sixth century, following the wanton destruction of the drinking water and sewage systems built by the Romans. Cats are themselves susceptible to bubonic plague, so they must also have died in large numbers, but evidently many survived.

Despite their self-evident usefulness and the considerable monetary value placed upon them, cats' welfare was not given the same respect it is today. In Greece, sacrifice and mummification of cats continued much as it had in Egypt, except that drowning rather than strangulation seems to have become the preferred method of killing. An ancient Celtic tradition of burying or killing cats to bring good luck spread across Europe, and given the value evidently placed on female cats, the usual victims were likely males. A cat would be killed and buried in a newly sown field to ensure the growth of the crop. A new house could be protected from mice and rats by putting a cat—whether dead or alive at the time of interment is unclear—and a rat into a specially constructed hole in an outside wall, or by placing them together under new floorboards.[2] Many European cities had a feast-day custom of putting several cats in a basket together—which would have been stressful enough in itself—and then suspending it over a fire; the screams of the cats were supposed to ward off evil spirits. An alternative was to throw cats from the top of a tower, an event the citizens of Ypres still commemorate every May as the Kattenstoet (Festival of the Cats), using stuffed toys as substitutes for actual cats.

Remarkably, some of these rituals persisted into modern times in their original form. In 1648, Louis XIV presided over one of the last cat-burnings in Paris, lighting the bonfire himself and then dancing in front of it before leaving for a private banquet. The last time live cats were thrown from the bell tower in Ypres was as recently as 1817. However repugnant such rituals may seem today, they cannot have accounted for more than a tiny fraction of the cat population; overall, cats probably thrived during the Dark Ages. Indeed, whenever mice were plentiful, cats would have bred prolifically, producing a surplus of kittens that would need to be culled, often by drowning, to prevent overpopulation. Cats in general were regarded as disposable, like any other farm animal: catskins were commonly used in clothing, and butchery marks on cat bones recovered from medieval sites show that many were killed specifically for their fur as soon as they were fully grown. Such practices would presumably have contributed to a prevailing notion that cats' lives were disposable, making their ritual sacrifice much less abhorrent then than it would be today.

Initially benevolent, the attitude of the Church towards cats gradually became more hostile as the Dark Ages gave way to the Middle Ages. The Roman Catholic Church first set itself on a road that ended in the wholesale persecution of cats, when in 391 CE the emperor Theodosius I banned all pagan (and "heretical" Christian) worship, including all the cults devoted to Bastet and Diana. Initially the worshipers themselves, not their cats, were targeted. Indeed, cats seem to have been favored animals in the early Irish Church. The Book of Kells, an eighth-century Irish illuminated book of the Gospels, features several illustrations of cats, some demonic and others depicted in domestic settings. Irish clerics may have encouraged an appreciation of cats as companions. The poem "Pangur Bán," by a ninth-century monk, compared the writer's life to that of his cat:

> *Myself and Pangur, cat and sage*
> *Go each about our business;*
> *I harass my beloved page,*
> *He his mouse. . . .*
> *And his delight when his claws*
> *Close on his prey*
> *Equals mine when sudden clues*
> *Light my way.*[3]

In the Middle Ages, monasteries must have been particularly valuable to cats because of their fishponds, used to raise the fish eaten during Lent, the liturgical season of the late winter and early spring when meat-eating was forbidden. Fish were plentiful and great sources of protein, and livestock were slaughtered months before because little fodder was available to feed them in the winter months. Cats, often pregnant during these months, would have been happy to turn from hunting mice to scavenging fish scraps, thereby gaining nutrition essential to the well-being of their unborn kittens, and an advantage over their heathen counterparts on nearby farms.

The relationship between Church and cat became seriously antagonistic from the thirteenth to seventeenth centuries, threatening the very survival of the domestic species in some parts of continental Europe. In 1233, the Catholic Church began a concerted attempt to exterminate cats from continental Europe. On June 13 of that year, Pope Gregory IX's notorious *Vox in Rama* was published. In this papal Bull, cats—especially black cats—were identified specifically with Satan. Over the next 300 years, millions of cats were tortured and killed, along with hundreds of thousands of their mainly female owners, who were suspected of witchcraft. Urban populations of cats were decimated. The justification for this barbarism was essentially the same as it had been in the fourth century—the extermination of cults that still included cats in their worship, and the demonization of rival religions such as Islam—but now it was the cats themselves that bore the brunt of the Church's wrath.

Outside Western Europe, cats were generally better tolerated at this time. The Eastern (Greek) Orthodox Church appears to have had little quarrel with cat-keeping. Islam has a tradition of kindness to cats, so cats continued to thrive in the Middle East. In Cairo, the Sultan Baibars, ruler of Egypt and Syria, founded what was probably the first sanctuary for homeless cats in 1280 CE.

Even within the reach of the Church of Rome, cats were not universally reviled. In Britain, cats feature in fourteenth-century poetry, including Geoffrey Chaucer's *Canterbury Tales*. If the "Steward's Tale" is accurate, then cats were well tended, as well as appreciated for their mouse-catching skills:

> *Let's take a cat and raise him well with milk*
> *And tender meat, and make his couch of silk,*
> *Then let him see a mouse go by the wall—*
> *At once he'll leave the milk and meat and all,*
> *And every dainty that is in the house,*
> *Such appetite he has to eat a mouse.*[4]

Moreover, it appears that cats were secretly popular even in ecclesiastical circles. Carvings of cats adorn the choir stalls of medieval churches across Europe, including Britain, France, Switzerland, Belgium, Germany, and Spain. These cats are depicted not as demons, but in natural or domestic situations—washing themselves, caring for kittens, and sitting by the fireside. It may be that such carvings were deliberately placed out of view of the general congregation, since presumably sermons sometimes referred to cats as demonic. Although cats and the women who doted on them were sporadically persecuted, cats were generally tolerated for their usefulness—not least in rural areas, where the reach of the Church was weakest and the services cats provided were most appreciated.

Did this reversal in its fortunes have any lasting impact on the cat as a domesticated animal? No physical trace of the persecution of black cats in Western Europe remains: today, the black mutation is as common in Germany and France as it is in Greece, Israel, or North Africa, all of which were outside the influence of the medieval Catholic Church. Sometime during the Middle Ages, cats, which were significantly larger and more wildcat-like up until Roman times, became smaller—in some locations, even smaller than the average cat of today. Although persecution might have influenced this, it is difficult to pinpoint the time or place that the changes in size occurred. In part this is because it is rare for sufficiently large numbers of cat bones to be found in one place to give an accurate picture of what the "average cat" might have looked like: for example, researchers would presume that the remains of a single large cat found anywhere in Europe are those of a wildcat, and an extra-small cat the product of malnutrition.

In Western Europe, the cat seems to have changed in size over the centuries, but not consistently. For example, in York in the tenth and eleventh centuries, domestic cats were the same size they are today, but at the same period in Lincoln, a scant eighty miles away, cats were mostly small by today's standards. However, by the twelfth and thirteenth centuries, York cats had become smaller than their counterparts of 200 years earlier. In Hedeby, Germany, cats of the ninth through eleventh centuries were roughly the same size as they are today, but in Schleswig, also in Germany, some of the cats recovered from eleventh- through fourteenth-century remains were tiny. A staggering and unexplained 70-percent reduction in bone lengths seems to have occurred between the eleventh and fourteenth centuries: many of the cats found in Schleswig, which seem to have left no descendants, would look tiny compared to a typical twenty-first-century cat.[5] We might be tempted to ascribe this miniaturization to the persecution that began in the fourteenth century, but we have no direct evidence that this was the case; indeed, the reduction in size in England predates the Papal proclamation. As such, the cause of these shifts toward smaller cats remains a mystery, and we do not know when cats grew larger again.

Likewise, we might be tempted to link the persecution of cats to the Black Death, a pandemic of rat-borne bubonic plague that swept from China to Britain from 1340 through 1350. More than a third of Europe's human population died, along with many of its cats. However, the plague was just as devastating in India, the Middle East, and North Africa, where cats remained unpersecuted, as it was in Western Europe. Evidently the bacillus was simply too virulent to be contained, and indeed plague continued to break out in Europe occasionally over the next 500 years. The last major epidemic in Britain was the Great Plague of London of 1665–66, and this time cats, not rats, were blamed; 200,000 cats were slaughtered on the Lord Mayor of London's orders.[6]

Seventeenth-century Britain was not a good time and place to be a cat, nor were the new colonies in North America. As a result of their link to the remnants of paganism in rural communities, cats—again, especially black cats—had become associated with witchcraft. We see remnants of this association today in horror films and Halloween decorations. When communities tried witches for their "crimes," they often claimed that the witches could transform themselves into cats; they also mentioned other animals, including dogs, moles, and frogs, but cats were the most common. Thus, the Church of Rome gave its official sanction to cruelty toward cats. Anyone coming upon a cat after dark was justified in killing or maiming it, on the grounds that it might be a witch in disguise. On the Scottish island of Mull, one black cat after another was roasted alive for four days and four nights in an exorcism referred to as the Taigherm.[7] Colonial leaders brought these same prejudices to Massachusetts, culminating in the Salem Witch Trials of 1692–93.

Witches with their "familiars"—a cat, a rat, and an owl

The reputation of the cat began to improve in mid-eighteenth-century Europe. Louis XV, the great-grandson of the same Louis XIV who had lit a Paris bonfire under a basket of cats, was at least tolerant of them, allowing his wife Maria and her courtiers to indulge their pet cats' every whim. It became fashionable for such pets to be incorporated into paintings of French ladies of title, and special tombs were built for favored animals when they died. However, such attitudes were far from universal: around the same time, the naturalist Georges Buffon wrote, in his definitive thirty-six-volume *Histoire Naturelle, Générale et Particulière*: "The cat is an unfaithful domestic, and kept only through necessity to oppose to another domestic which incommodes us still more, and which we cannot drive away."[8]

Meanwhile in England, cats were growing in popularity. Eighteenth-century poets Christopher Smart and Samuel Johnson not only valued the company of cats, but also wrote about them. Smart's poem *For I Will Consider My Cat Jeoffry* begins:

> *For I will consider my Cat Jeoffry.*
> *For he is the servant of the Living God duly and daily serving him.*
> *For at the first glance of the glory of God in the East he worships in his way.*
> *For this is done by wreathing his body seven times round with elegant quickness.*

—implying that Smart saw no link between cats and devil worship: indeed, quite the opposite. Smart was also an acute observer of cat behavior. The poem continues:

> *For first he looks upon his forepaws to see if they are clean.*
> *For secondly he kicks up behind to clear away there.*
> *For thirdly he works it upon stretch with the forepaws extended.*
> *For fourthly he sharpens his paws by wood.*
> *For fifthly he washes himself.*

For sixthly he rolls upon wash.
For seventhly he fleas himself, that he may not be
interrupted upon the beat.
For eighthly he rubs himself against a post.
For ninthly he looks up for his instructions.
For tenthly he goes in quest of food.

Likewise, Samuel Johnson adored his cats Hodge and Lily. His biographer James Boswell wrote, "I never shall forget the indulgence with which he treated Hodge, his cat," possibly referring to Johnson's habit of feeding Hodge oysters—which, it must be said, were not the luxury food then that they are now.

Toward the end of the nineteenth century, the cat completed its transformation to domestic status. In Britain, Queen Victoria kept a succession of pet cats: her Angora, White Heather, was one of the consolations of her old age, surviving her to become the pet of her son, Edward VII. In the United States, Mark Twain was not only a cat enthusiast but also, like Smart, an acute observer of their nature:

> By what right has the dog come to be regarded as a "noble" animal? The more brutal and cruel and unjust you are to him the more your fawning and adoring slave he becomes; whereas, if you shamefully misuse a cat once she will always maintain a dignified reserve toward you afterward—you will never get her full confidence again.

By the nineteenth century, cats became far more varied in appearance than their ancestors in Egypt. At different times and in various places in Europe and the Middle East, new coat types appeared due to spontaneous genetic mutations. Sometimes these must have disappeared after a few generations, particularly if cat owners regarded the mutations as freakish. However, sometimes unusual-looking cats must have found sympathetic owners who treasured these unique differences. If this preference came to be shared by local owners, and then spread to other places, the underlying mutation might gradually spread through the general population, and thereby became part of the range of variation that we see in today's cats—different colors, different patterns, both long and short hair. Remarkably, we can still trace the origins and spread of some of these changes, even in today's multifarious cat populations.

Geneticists credit the comparative homogeneity of household cats across Europe and the Middle East to the local habit of keeping cats on ships. For example, we can easily imagine a litter of kittens conceived in Lebanon being born two months later in Marseilles after their mother had jumped ship. Seagoing Phoenicians, Greeks, and Romans spread cats around the Mediterranean, and the genetics of cats in France still shows traces of the trade route up the river Rhône from the Mediterranean and then down the Seine to the English Channel. The distribution of the orange tabby mutation in northern Europe still shows the effects of its popularity with Viking invaders almost a thousand years ago.[9]

The first evidence for domestic cats that did not have striped coats comes not long after the beginning of the Christian era. The Greeks first described black cats and white cats in the sixth century CE, but the black mutation, which occurs quite regularly, may have been spreading through the domestic population for several centuries before. Many of the cats that carry this mutation are not actually black. The black color is caused by an inherited inability to produce the normal pale hair tips that give wildcats their brownish appearance, which is known technically as "agouti." The hairs then revert to their basic color, black, but only if the cat carries two copies of this mutation—one inherited from its mother and one from its father. If it carries one normal copy and one mutant, then the normal copy is dominant and its coat will be the usual tabby. Thus, a male and female pair of such cats, outwardly tabby, could produce a black kitten alongside their other, tabby, kittens. The black mutation is widespread among today's cats, wherever in the world they are found, suggesting that it may have originated before the Phoenicians and Greeks began spreading domestic cats around Europe.

Black cats and white cats are not only opposites in terms of color; they are also opposites in terms of their relationship with humankind.

Cats can be all white either because they are albino, in which case their eyes will be pink, or because they carry a mutation, "dominant white." Both tend to be less healthy than normal cats; not only are they prone to skin cancer, but blue-eyed "dominant white" cats are often deaf. Perhaps more importantly, unlike their striped counterparts, all-white cats stand out against almost any background, so are unlikely to find it easy to catch enough food to sustain themselves. Outside the pedigree population, in which dominant white coats have been deliberately bred into some breeds, all-white cats are uncommon, rarely forming more than 3 percent of the free-ranging cat population.

By contrast, the black mutation is so common and widespread that it must not bring any major biological disadvantage to the cats that carry it. In some places, more than 80 percent of cats carry this mutation—not that all or even most of these cats will actually have black coats, since many will have only one copy of the mutation and will therefore appear striped. Today, black is commonest in Britain and Ireland, in Utrecht in the Netherlands, in the city of Chiang Mai in northern Thailand, in a few US cities such as Denton, Texas (holding the current record at almost 90 percent incorporation), in Vancouver, and also in Morocco.

Since black has not always been everyone's favorite color for a cat, this ubiquity is difficult to fathom. Since cultural explanations appear not to fit, scientists have proposed a biological basis for the pervasiveness of the black mutation: that possession of this mutation, even one invisible copy (the "heterozygote"), somehow makes the cat more friendly to people and/or other cats, thereby giving the cats that carry it an advantage in high-density living situations or in prolonged, unavoidable contact with people, such as onboard a ship.[10]

This hypothesis runs counter to a recent study of cats in Latin America.[11] Here, roughly 72 percent of cats carry the black mutation, similar to the proportion in Spain, from where most South American cats must have come. We can estimate the number of long journeys that cats must have taken between Spain and each of the various Hispanic colonies by tracing their paths along the trade routes. The ancestors of the cats of La Paz, 13,000 feet above sea level in the Andes,

Blotched and striped tabbies

completed many such journeys, but black cats are no more common here than in Spain; any cat that traveled this far must have been remarkably tolerant of people. Thus, we cannot yet convincingly explain either the overall numbers of black cats or their local variations.

The "classic" or blotched tabby pattern exhibits the most remarkable distribution, although the reasons for its prevalence in some areas and not in others are not entirely clear. The cat's wild ancestor had a striped, or "mackerel," tabby coat, and all domestic cats probably carried the genes for this pattern until at least 2,000 years ago, and possibly even more recently.[12] The mutation for the blotched tabby pattern probably first took hold sometime in the late Middle Ages, and almost certainly in Britain, where it is the commonest pattern today.

Like the black mutation, blotched tabby is recessive: for a cat to have a blotched tabby coat, it must have two copies of the blotched version of the gene, one inherited from its mother and one from its father. One striped, one blotched, and the cat will have a striped coat. Despite this apparent handicap, in Britain and in many parts of the United States, blotched tabbies outnumber striped tabbies about two to one, meaning that more than 80 percent of cats carry the blotched version of the gene. In many parts of Asia, blotched tabbies are rare or even absent. The main exceptions are a few former British colonies such as Hong Kong, which were presumably simultaneously colonized by British cats, either ships' cats or the pets of the settlers.

For a new version of a gene, especially a recessive gene, to spread through a population, it must provide some advantage. Since (striped) cats had been in Britain since Roman times and probably earlier, blotched tabbies must have been rare to begin with. They probably reached 10 percent of the population by about 1500 CE, and then increased year after year until they reached their current near-ubiquity. In Britain, blotched tabby is sometimes referred to as "classic" tabby, as if the striped version were the mutation, not the other way around.

The reason for the ascendancy of the blotched pattern is still unknown. British cat owners do not prefer the blotched to the striped coat, at least not to the extent that this could account for their proportions among British pets; in fact, when asked, they express a small preference for striped tabby, perhaps simply because it is (now) relatively unusual. Blotched tabby does not seem to provide any better or worse camouflage than striped, at least in the countryside. There has been a suggestion that the pollution caused by the Industrial Revolution, coating British cities in soot, favored darker cats—both blotched tabby and black—because they were less conspicuous, but this has never been confirmed.[13] Still, we know that almost all genes have multiple effects, even though most are named for the most obvious change they bring about. Therefore, the blotched version of the gene possibly produces some other advantage, nothing to do with the coat, that somehow suited cats to life in Britain.

We see the rise of the blotched tabby gene in Britain reflected in the proportion of blotched tabbies in former British colonies all around the world. In the Northeastern United States—New York, Philadelphia, and Boston—settled in by Europeans in the 1650s, only about 45 percent of cats carry the blotched tabby gene, but this is considerably more than in originally Spanish-settled areas, such as Texas, at around 30 percent—where cats look much like those in Spain today. As shown in the nearby figure, the Atlantic Provinces of Canada, settled some 100 years later, have more blotched tabbies. European colonies settled in the nineteenth century are more variable: Hong Kong in particular has fewer than it should, probably because there was already a striped-tabby population of Chinese origin there, thus diluting the effect of British immigration. Australia on the other hand has more than it should, possibly the result of later waves of British immigrants in the twentieth century bringing their cats with them. The proportion in Britain was over 80 percent in the 1970s, and may have continued to rise since then.

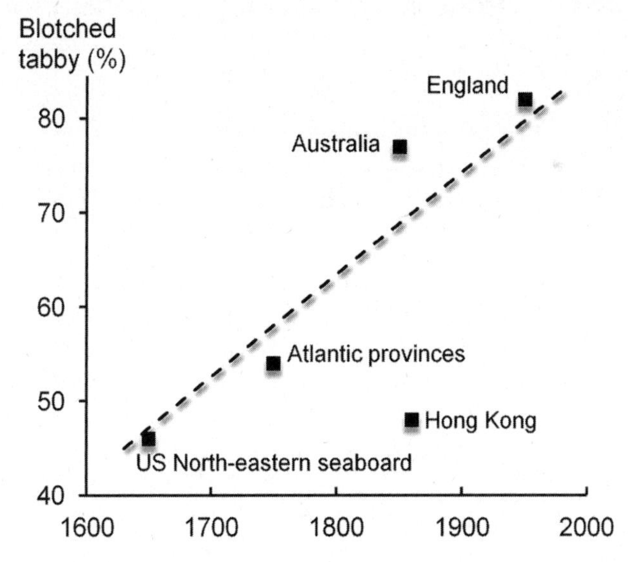

How the percentage of the blotched tabby allele varied between locations colonized from Britain between 1650 and 1900, compared to England in 1950

We can explain this trend by making two assumptions. The first is that the proportion of blotched tabbies in Britain has been rising steadily since about 1500, for some reason unique to that country—otherwise the same change should have taken place in, for example, New York, Nova Scotia, Brisbane, and Hong Kong, resulting in blotched tabbies reaching 80 percent of the population today in those places as well. The second is that once a population of cats becomes established in a particular place, the proportion of blotched to striped doesn't change. The latter assumption holds for other places and variations in coat color, too, so it may be universal (see box below, "Reconstructing the Origins of the Cats of Humboldt County, California"): however, given that it seems not to have occurred anywhere else, the rise of the blotched tabby in Britain becomes even more curious.

Reconstructing the Origins of the Cats of Humboldt County, California

Domestic cats arrived on the West Coast of the USA by a variety of routes—by sea from the south and the north, and overland from the east. From the sixteenth to eighteenth centuries, Humboldt County, on northern California's Redwood Coast, received a succession of visitors and settlers, including Russian, British, and Spanish vessels exploring the Pacific Coast, and farmers from both Missouri to the east and Oregon to the north. Feral cats, probably escapees from trading ships, were first recorded in Humboldt County in the 1820s, before the arrival of the first farmers, so today's cats could be the descendants of either—or both.

In the 1970s, biologist Bennett Blumenberg recorded the coat colors and patterns of 250 local cats, and from these determined the proportions of the different versions of each gene. For example, 56 percent of the cats were black or black and white, from which he was able to estimate that the black ("non-agouti") version of the gene was present in 75 percent of the population—the difference being made up of cats that were carrying one black version and one tabby, and were therefore outwardly tabby. He also recorded the numbers of orange and

(continues)

Reconstructing the Origins of the Cats of Humboldt County, California *(continued)*

tortoiseshell cats, blotched vs. striped tabbies, pale coats, long and short hair, all-white cats, and cats with white feet and bibs: each one of these variations is controlled by a different and well-understood gene. He then compared these with cats from other parts of North America.[14]

The most similar were the cats of San Francisco, Calgary, and Boston, indicating that the ancestors of today's Humboldt County cats had mainly arrived overland with farmers and prospectors, or in the fur-traders' vessels sailing out of Boston. However, he also found traces of cats of Spanish origin, left behind after the demise of New Spain and its occupation of much of what is now California. Some otherwise unaccounted for similarities were detectable with the cats of Vladivostok, homeport of the Russian-American Trading Company. Their ships also sailed from ports that are now in China, and could feasibly have carried cats of Chinese origin, but Blumenberg found no trace of Chinese cat genes in the Californian cats.

Polydactyl footprints

Another rare mutation known as polydactyly gives cats an extra toe on each foot. Early in the establishment of Boston, one newly arrived cat must have produced a kitten with extra toes, and that kitten became the ancestor of many more, such that by 1848, extra-toed cats were common there; today, they form about 15 percent of the population. Cats with extra toes are also common in Yarmouth, Nova Scotia, a seaport founded by immigrants from Boston, whereas in nearby Digby, settled by New York Loyalists at the end of the Civil War, polydactyly is as rare as it is everywhere else.[15]

Relatively few of the domestic cats that colonized South America from Spain and Portugal were blotched tabbies, so scientists have had to turn to other variations in appearance to trace their history. Here also, it seems that once a population of cats becomes established, its genetics doesn't change much, even over several centuries. It seems plausible that the cat population is set up when the initial (human) colonists bring their cats with them, and subsequent immigrants only adopt cats from the local population. For example, in the nineteenth century, Catalans from Barcelona established several towns along the river Amazon, and the cats of at least two of those towns, Leticia-Tabatinga and Manaus, remain similar to the cats of Barcelona more than a century later.[16]

As in Hispanic-settled parts of the United States, and in Spain itself, orange and tortoiseshell cats are more common in South America than in most other places. The exceptions, where these colors are even more common, are Egypt—possibly one origin of this mutation (see the box on page 41, "Why Ginger Cats Are Usually Boys"); the islands off the north and west coasts of Scotland; and Iceland. Researchers have attributed these exceptions to a hypothetical preference for orange cats among the Vikings, who colonized those places around the ninth century CE, but it is unclear whether the Vikings first obtained their orange cats from the eastern Mediterranean, or whether orange appeared spontaneously and independently somewhere in Norway and was then transported around the Norwegian Sea in Viking ships.

Today, cats come in many colors and coat types, and many of these variations are undoubtedly the result of human preferences—even leaving aside pedigree cats, whose breeding is strictly under human control. The basic colors of cats—black, tabby (striped or blotched), ginger—are well established in populations all around the world, albeit in slightly different proportions. These colors persist in cats that have gone feral, so they do not seem to produce any major advantage—or disadvantage—to their wearers. However, many other variations in appearance seem to persist mainly because people like them. Cats with white feet and variable amounts of white on the body (the gene that controls this is rather imprecise in its effects) are less well-camouflaged than their plainer counterparts, putting the cat at a disadvantage when hunting. However, many people prefer their cats to carry white patches, especially the black-and-white "tuxedo" cat. Some people also favor cats that carry a gene that dilutes their coat color, such that black cats become a fetching shade of gray (often referred to as "blue"), and other colors are a few shades lighter than usual.

Some people like long-haired cats. Although it seems self-evident that these cats, which like all cats cannot sweat through their coats, should be at a natural disadvantage in warm climates, a recent study of Latin American cats indicates that human preference is a much more potent factor than the weather, although a thick coat, such as that of the Maine Coon and the Norwegian Forest Cat, is undoubtedly beneficial for outdoor cats in cold climates.[17] In temperate climates, the main disadvantage of long hair is not that the cat overheats, but that its coat easily becomes matted, which, if left unattended, leads to infection or infestation of the skin. Long-haired cats are rarely seen in feral colonies, testifying to their unsuitability for a life without attention from humans.

The striped tabby pattern evidently suits wildcats the best, possibly because it provides the most effective camouflage for hunting. Presumably, mutations that affect appearance have occurred from time to time, but the wildcats that happened to carry them fared worse than their "normal" counterparts, so the mutation quickly died out. For a domestic cat, camouflage is probably less critical, allowing other coat colors to spread through the population.

A similar variety of colors and coat types has, of course, also become a feature of many other types of domesticated animal—dogs, horses, and cattle, for example. In the case of the cat, external appearance reflects two distinct factors, both of which affect how many offspring a cat is likely to have, and therefore how common its coat color and type are likely to be in the next generation. The first is whether these factors impede the cat's hunting ability; this may not be so important today, but it certainly was in the past. The second is how appealing the cat is to its owner. Because human tastes vary from person to person and culture to culture, this second factor has produced many variations.

What seems to be missing here is any trace of each cat's own preferences. Although never studied directly, we have no evidence for "color prejudice" among cats. Tabby females do not seem to spurn black toms in favor of other tabbies: a white bib and socks seems to be neither an advantage nor a disadvantage when it comes to finding a mate, except perhaps indirectly, if the cat's conspicuousness has resulted in it catching less prey and thus looking less healthy than its better-camouflaged rival. Whatever criteria cats use to choose a mate, coat color seems to be well down the list.

Even today, most domestic cats exert a remarkable amount of control over their own lives, significantly more than other domestic animals, such as dogs. If we leave the pedigree breeds to one side for a moment (and these are still in a minority), most cats go where they please and choose their own mates (unless they are neutered—a relatively recent phenomenon). For this reason alone, cats cannot be considered completely domesticated.[18] Full domestication means that humankind has complete control over what an animal eats, where it goes, and most crucially, which individuals are allowed to breed and which are not.

We certainly provide domestic cats with most or all of their food, but in this respect, too, cats are an anomaly. Science classifies them as obligate carnivores, animals that have to obtain a mostly meat-based diet if they are to thrive (see box on page 71, "Cats Are the True Carnivores"). For a female cat to breed successfully, she must have a high proportion of flesh in her diet, especially in late winter, when she is preparing to come into season, and subsequently while she is pregnant. The domestication of such an animal takes some explaining; until recently, meat and even fish formed only a minor part of most people's diets, and were often only seasonally available. Most domesticated animals will thrive on foods that we ourselves cannot eat, unlocking sources of nourishment that would otherwise be unavailable to us: cows turn grass, which we can't digest, into milk and

Cats Are the True Carnivores

Cats are carnivores not by choice, but by necessity. Many of their relatives in the animal kingdom, though referred to as Carnivora, are actually omnivores—including domestic dogs, foxes, and bears—and some, like pandas, have reverted to being vegetarian. The whole cat family, from the lion down to the tiny black-footed cat from southern Africa, have the same nutritional needs. At some point many millions of years ago, the ancestral cat became such a specialized meat-eater that it lost the ability to live on plants: it became a "hypercarnivore." Once lost, such capabilities rarely re-evolve. Domestic cats might have been more successful if they could have gotten by, as dogs can, living on scraps, but they are stuck firmly in the nutritional dead end their ancestors bequeathed to them.

Cats require far more protein in their diet than dogs or humans do, because they get most of their energy not from carbohydrates but from protein. Other animals, faced with a shortage of protein in the diet, can channel all the protein they do get into maintaining and repairing their bodies, but cats cannot. Cats also need particular types of protein, especially those that contain the amino acid taurine, a component that occurs naturally in humans, but not in cats.

Cats can digest and metabolize fats, some of which must come from animal sources so that the cat can use them to make prostaglandins, a type of hormone essential for successful reproduction. Most other mammals can make prostaglandins from plant oils, but cats cannot. Female cats must get enough animal fat during the winter to be ready for their normal reproductive cycle, mating in late winter and giving birth in the spring.

Cats' vitamin requirements are also more stringent than ours. They need vitamin A in their diet (if need be, we can make ours from plant sources), sunshine doesn't stimulate their skin to make vitamin D as ours does, and they need lots of the B vitamins niacin and thiamine.[19]

None of this is a problem if the cat can get plenty of meat—although raw fish, which contains an enzyme that destroys thiamine, can cause a deficiency if eaten in excess. It is slightly possible to construct a vegetarian diet for cats, but only if every single one of the cat's nutritional peculiarities are carefully compensated for. Their taste buds also differ substantially from our own, having evolved to focus better on an all-meat diet. They cannot taste sugars; instead, they are much

(continues)

Cats Are the True Carnivores *(continued)*

more sensitive than we are to how "sweet" some kinds of flesh are, compared to others, which they find bitter.

Cats do have two notable nutritional advantages over humans. First, their kidneys are very efficient, as expected for an animal whose ancestors lived on the edge of deserts, and many cats drink little water, getting all the moisture they need from the meat they eat. Second, cats do not require vitamin C. Taken together, these make cats well suited to shipboard life: they don't compete with sailors for precious drinking water, getting all they need from the mice they catch, and they are not afflicted by scurvy, a common disease among mariners to the middle of the eighteenth century, when it was found it could be prevented by eating citrus fruits.

meat, which we can. (Though classified as carnivores, even dogs are actually omnivores; they may prefer meat, but cereal-based foods can, if necessary, give them all the nutrition they need.)

For most of their coexistence with humankind, cats were valued primarily for their skill as hunters. Since mice contain all the nourishment a cat needs, a successful hunter automatically ate a balanced diet; starvation was always possible for the less adept or the unlucky, but diseases due to specific nutritional deficiencies were unlikely. However, even historically, few domestic cats would have lived by hunting alone, most being provided with at least some food by their owners, supplementing their diet by scavenging. So long as some of their diet came from fresh meat, usually in the form of prey they had killed themselves, scavenging would not have tipped them into nutritional imbalance; still, giving up hunting entirely would have been risky.

Cats do not scavenge at random; they have some ability to make informed choices. They subsequently avoid foods that make them feel seriously and instantly ill. When eating food that they have not killed for themselves, they also deliberately seek out a varied diet, thereby avoiding a buildup of anything that might make them sick long-term, but might slip beneath the radar of immediate malaise.

To demonstrate this behavior, I laid out individual pieces of dry cat food on a grid on the ground, some of one brand, some of another, and then allowed rescued strays to forage over them, one at a time. In this way, I could record in precisely what order each cat picked up and ate the pieces of the two foods. When there were equal numbers of each type of food, the cats roamed across the grid, eating both foods but more of whichever food they liked better. However, when one of the two foods made up 90 percent of the total, every cat, whatever its preference, stopped grazing indiscriminately within a couple of minutes and started actively seeking out the rarer food. Thus, these cats demonstrated a primitive "nutritional wisdom," as if they assumed that eating a variety of food was more likely to produce a balanced diet than simply eating the food that was easiest to find (even though both the foods offered were nutritionally complete).[20] When I gave similar choices to pet cats that had always had a balanced diet, few responded this way, most continuing to eat whichever of the two foods they liked better from the outset, or was just easy to find. Thus, although all cats probably have the capacity to deliberately vary their diets, it seems that this ability must be "awakened" by some experience of having to scavenge for a living, as most of the stray cats in our original experiment must have done before they were rescued.[21]

Most other animals have more varied diets than cats'. Rats, the best-studied example, are omnivores with extremely wide tastes that suit them perfectly for a scavenging lifestyle. They employ several strategies that enable them to pick the right foods from the wide but unpredictable choices available to them. New foods are only nibbled at until the rat is sure that they're not poisonous. As soon as each food starts to be digested, the rat's gut sends information to its brain about its energy, protein, and fat content, enabling the rat to switch to another food with a different nutritional content if necessary. Cats are much less sophisticated in this regard, having traveled a different evolutionary path based on mainly eating fresh prey, which is by definition nutritionally balanced.

Given their limited ability to subsist by scavenging alone, cats were locked into the hunting lifestyle until as recently as the 1980s. Until

Cats foraging on a grid

science revealed all of their nutritional peculiarities, it would have been a matter of luck whether a cat that wasn't able to hunt obtained a nutritionally balanced diet, unless its owner was both willing and able to give it fresh meat and fish every day. Although commercial cat foods have been available for over a century, there was initially little understanding that cats had very different requirements to dogs, and much of this food must have been nutritionally unbalanced. Commercial cat food that is guaranteed to be nutritionally complete has been widely available for only thirty-five years or so—only 1 percent of the total time since domestication began.

In evolutionary terms, this is just a blink of an eye, and we have yet to see the full effects of this improvement in nutrition on cats' lifestyles. Just a few dozen generations ago, the cat that was a skillful and successful hunter was also the cat that stood the best chance of breeding successfully. Those cats that depended entirely on man for their food would usually obtain enough calories to keep them going day-to-day, but many would not have bred successfully, because many of the cat's unusual nutrient requirements are essential for reproduction. Nowadays, any cat owner can go to the supermarket and buy food that keeps their cat in optimum condition for breeding. That is, of course, if the cat has not been neutered—another development that has yet to show its full effect on the cat's nature.

Today's cat is thereby a product of historical turmoil and misconception. What would today's cats be like if they had not gone through centuries of persecution? It is possible that the effects may not have been particularly long-lasting. After all, since serious attempts were made to exterminate black cats in continental Europe because of their supposed association with witchcraft, they should still be uncommon there today—and they're not. Although undoubtedly many individual cats did suffer horribly, little lasting damage seems to have been done to the species as a whole. This is probably because, for most of the time and in most places, particularly in the countryside, cat-keeping was both enjoyable and practically beneficial, even if occasionally interrupted by an outburst of religious persecution. Changes have taken place—cats today are significantly smaller and more varied in color than when they left Egypt—but these changes appear to have been mainly local rather than global.

Thus, following the origin of their partnership in ancient Egypt, cats and humans continued to live alongside one another, for a further 2,000 years, without the cat ever becoming fully domesticated. Then, due to the nutritional discoveries of the 1970s, all cats, and not just the pets of the well-off, were relieved of the necessity to hunt for a living. However, their predatory past, so essential to their survival until recently, cannot be obliterated overnight. One of the most significant challenges facing today's cat enthusiasts is how to allow their cats to express their hunting instincts without causing wildlife the damage which evokes so much criticism from the anti-cat lobby.

CHAPTER 4

Every Cat Has to Learn to Be Domestic

Cats are not born attached to people; they're born ready to learn how to attach themselves to people. Any kitten denied experience with people will revert toward its ancestral wild state and become feral. Something in their evolution has given domestic cats the inclination—and it's no more than that—to trust people during a brief period when they're tiny kittens. This minute advantage enabled a few wildcats to leave their origins behind and find their place in environments created by the planet's dominant species. Only one other animal has done this more successfully than the domestic cat, and that, of course, is the domestic dog.[1] Like puppies, kittens arrive into the world helpless, and then have just a few weeks in which to learn about the animals around them—an even shorter time for cats than for dogs—before they must make their own way in the world. By comparison with our own infants, which are dependent on us for years, this is a very brief period. Even in their wild ancestors, the wolf and the wildcat, this window must have been open just a crack, waiting for evolution to allow the young animals of these two species to learn to trust us, and thereby become domesticated.

Kittens and puppies alike become more closely integrated into human society than any other animal can, but the way the two species achieve this differ. Early scientific studies of dogs from the 1950s established the notion of a primary socialization period, a few weeks in the puppy's life when it is especially sensitive to learning how to interact with people. A puppy handled every day from seven to fourteen weeks of age will be friendly toward people and virtually indistinguishable from a puppy whose handling started four weeks earlier. For the next quarter century, scientists generally assumed that kittens must be the same, and that it was not essential to handle kittens until they were seven weeks old. In the 1980s, when researchers finally performed corresponding tests on cats, those recommendations had to change.

These experiments confirmed that the concept of a socialization period could indeed be applied to cats, but that this period was comparatively curtailed in kittens. The researchers handled some kittens from three weeks old, some from seven weeks old, and the rest not until the testing started at fourteen weeks. The kittens started learning about people much earlier than puppies do. As expected, the kittens handled from their third week were happy to sit on a lap when they reached fourteen weeks old, but those whose contact with people had been delayed to seven weeks jumped off within half a minute—though not as quickly as those that had never been handled during their first fourteen weeks, which typically stayed put for less than fifteen seconds.

Handling a feral kitten

Could this be explained by the seven-week-handled kittens being more active than the three-week—in other words, no less happy to be on a person's lap, just more eager to explore their surroundings? It quickly became obvious that this could not be the cause. When each kitten was subsequently given the opportunity to cross a room toward one of their handlers, only the three-week kittens did so reliably—and were quick to do it, too, giving every impression that they were attracted to the person, who was by then very familiar to them. The seven-week and unhandled kittens did not seem unduly frightened of the person, and would occasionally get close to her. Some even apparently asked to be picked up, but these two groups were more or less indistinguishable in their behavior.

The handling that those seven-week kittens received up to the point of testing had not produced the powerful attraction to people that was obvious from the behavior of the three-week kittens. For the scientists taking part, the tests simply formalized what was already obvious from the kittens' behavior. As the leader of the research team noted, "In observing and interacting with these cats during testing and in their home rooms, it was obvious to everyone working in the lab that the late-handled [i.e., seven-week] cats behaved more like the unhandled cats."[2]

The scientists concluded that cats need to start learning about people much earlier than dogs must. Dog breeders ought to handle puppies before they are eight weeks old, but if puppies are not handled until that age, then with the right remedial treatment they can still become perfectly happy pets. A kitten that encounters its first human in its ninth week is likely to be anxious when near people for the rest of its life. The paths that lead to an affectionate pet on one hand and a wild scavenger on the other diverge early in the cat's life; indeed, if it were any earlier, few cats would be able to forge relationships with us.

Even though the most crucial changes start in the third week, the first two weeks of the kitten's life are far from uneventful. For the first fourteen days of a kitten's life, the most important feature in its world is its mother. Kittens are born blind, deaf, and incapable of moving more than a few inches unaided. They cannot regulate their own body temperatures. Especially if the litter is larger than average, and the kittens are correspondingly small, they carry little in the way of energy reserves, and even the loss of a fraction of an ounce in weight can lead to weakness and, if outside intervention is not forthcoming, death. Their survival, then, depends on their mother's capabilities. Tomcats play no part in rearing offspring, and many mothers that give birth outside the home raise their kittens without help. The mother's choice of nest site is crucial, especially when the luxury of giving birth indoors is not available. Kittens must be well protected from the weather and from potential predators. After the first twenty-four hours, which all mother cats spend suckling and grooming their kittens, she may have to leave them to find food; and if that involves hunting, as it would in the wild, that may take some time.

Once kittens are a few days old, the mother may move them to a different nest site if her instincts make her feel uncomfortable about the original den. Kittens have a special "scruffing" reflex that enables her to do this quickly and quietly, without drawing attention to their vulnerability: she grasps them in her mouth by the loose skin on the back of their necks, and they instantly go limp and apparently oblivious to their surroundings, until she drops them in the new nest, where they appear confused but otherwise unaffected (see nearby box, "Clipnosis").

"Clipnosis"

Unlike much infantile behavior, in some cats the "scruff" reflex mother cats use when carrying their kittens persists into adulthood. For those individuals, scruffing can be used as a gentle and humane method of quieting a fearful cat. The cat is simply grasped firmly by the skin on the back of its neck and, if the reflex is triggered, may go into what appears to be a trance, enabling it to be picked up and carried; its weight must be supported by the other hand. Veterinary nurses sometimes use a hands-free version, applying a line of several clothespins to the area of skin between the top of the head and the shoulders. By doing so, the nurses can complete an examination of the cat without causing it too much stress.[3]

Many mother cats try to move their litters at least once before they wean them, but science has yet to find out why. In the wild, a mother cat will inevitably carry a few fleas, and because she has to spend so much time in the nest, flea eggs accumulate there; when they become adult fleas three or four weeks later, a single hop will take them onto a kitten. Removing the kittens from such an obvious source of infestation seems a good strategy and may be one explanation for nest-moving, but so far no evidence has been found that this reduces the number of fleas that kittens carry. Nest-moving may simply be a response to a disturbance that made the mother anxious, or may sometimes be strategic, bringing the kittens closer to the source of food onto which their mother intends to wean them. The kittens may suffer if the mother cat moves them too early or chooses the wrong new nest site. Kittens are vulnerable to becoming chilled, especially in damp weather that also favors the transmission of respiratory viruses; many feral kittens, particularly those born in the autumn, succumb to cat flu.

As my own experience shows, not all cats are innately skilled at parenthood. My cat Libby, having a nervous disposition, was not the best mother. As the time for her to have her litter approached, we tried to keep her in the same room where she had herself been born, presuming she would find the surroundings reassuring. But she was restless, patrolling the house and looking into every open cupboard and drawer, as if undecided as to the safest location. At least she showed no inclination to give birth outdoors. Eventually, she decided to produce her kittens within a few feet of where she herself had come into the world.

We should not have relaxed: a few days later, we discovered the kittens scattered all around the house. For a few hours after the birth, Libby lay with her three kittens and allowed them to suckle, but after that she appeared to lose interest, spending too much time away from them. Her mother Lucy was intrigued by the kittens, but at this stage played no part in looking after them. When we checked the kittens' weights, they seemed to be growing, so we were not unduly worried until Libby began to try to pick them up in her mouth and carry them away from the den we had built for her. Her inexperience as a first-time mother showed immediately, as she grasped the kittens rather roughly by the head, only occasionally and apparently accidentally gripping them correctly by their scruffs. Once she had gotten the hang of carrying them, she began to look for places to hide them.

Without our intervention, Libby's kittens would surely have perished. She would find a secluded place to hide the first kitten, and then return to the other two, pick up another, and take that one somewhere else, ignoring the cries of the first. After taking the third to yet another location, Libby would wander off as if unsure what to do next. Each time this happened, we searched out all the kittens and put them back in the original den. Once or twice, we tried constructing a new den—in the same room, but with completely new bedding so that it smelled nothing like the first, hoping this might fool Libby into believing she had successfully moved the whole litter herself. Still, the removals persisted. Finally, grandmother Lucy, her maternal instincts aroused, began to retrieve the kittens herself. Libby, perhaps sensing that she should follow her mother's lead, gradually gave up trying to move the litter. She continued to feed them, and they grew stronger, but from then on it was Lucy who groomed them and kept them together until they were ready to be weaned.

Lucy seemed to know the kittens by their appearance and by their smell, but first-time mothers can behave as if they do not immediately "know" what a kitten is. Rather, they respond to a single powerful stimulus that ensures that they take care of their newborns. This is the kitten's high-pitched distress call, uttered if it is cold, hungry, or out of contact with its littermates. When the mother cat hears this sound coming from outside the nest, indicating that a kitten has strayed away, she should instantly begin a search and, when she finds the kitten, retrieve it by its scruff. If the kittens are all in the nest and calling together, her instinct is to lie down and encircle them with her paws, drawing them onto her abdomen and allowing them to suckle. Gradually, over the course of their first couple of weeks of life, she seems to recognize the kittens as independent animals, though whether she ever comes to know them as individuals is unclear.

Libby's kittens preferred to curl up with Lucy

Kittens are so vulnerable in the first few weeks of their lives that their survival depends almost entirely on their mother's skill. Libby seemed to lack several components of maternal instinct, but even if just one is defective, the kittens' chances become slim. Yet if cats are as little removed from the wild as many researchers consider they are, they should—and most do—retain the ability to get everything right the first time. Research on free-ranging cats has produced little evidence to support the idea that first-time mothers are less successful than second-time mothers, although Libby's inability to locate her kittens' scruffs is reportedly quite common in inexperienced mother cats (known as queens).[4] Of course, domestication has provided an obvious safety net: the intervention of human owners.

For about the first two weeks of their lives, kittens define their world through smell and touch. At birth, their eyes and ears are still sealed, providing them with little useful information. Kittens recognize their mother immediately, initially just by her warmth and feel, and rapidly also learn her characteristic smell. They probably have little idea of what she is "supposed" to smell like; orphaned kittens fed on an artificial "mother" that smells nothing like a cat quickly learn its scent instead—a kind of "imprinting." In one classic series of experiments, researchers constructed a surrogate mother from artificial fur fabric, through which latex teats protruded, only some of which provided milk. Each teat was scented differently, for example with cologne or oil of wintergreen. The kittens quickly learned which scent indicated a milk-delivering teat.[5]

Each kitten gradually develops a nipple preference, based also on the nipple's odor rather than its position, since each kitten knows precisely where it wants to suckle whichever way its mother is lying. They find their preferred nipple by following the odor trail made up of their own saliva, and probably secretions from scent glands under their chins, left behind on their mother's fur as they approach the nipple.[6]

Kittens are unusually flexible when bonding with their mothers, a quality that works in their favor when their mothers are part of a social group. Often these groups are composed of close relatives, say a female and her adult daughters, which have grown up together and know one another well enough to overcome their natural mistrust of other cats. Such cats will spontaneously share nests and pool their kittens.

Local authorities once asked me how to handle such a cat colony living underneath some temporary buildings. Having pointed them toward humane alternatives to putting out poison, I located an animal charity to trap the cats and relocate them.[7] It was springtime, and three of the females were heavily pregnant; they gave birth within a few days of one another. Even though they were given three separate boxes, the mothers soon put all the kittens together, and each fed them indiscriminately. Ten kittens feeding from three queens, all within a few inches of one another, produced the loudest chorus of purring I have ever heard.

As this implies, mother cats can be equally indiscriminate when identifying their own kittens from those of other cats—or at least, those of their relatives, which may smell similar. For cats, the general rule appears to be that if a kitten is in your nest, it must be yours. Some cat rescue organizations exploit this quality to raise orphaned kittens: some mother cats will readily accept an extra kitten that is gently slipped in with their own, even if their ages don't match up. Some will even suckle entire litters introduced to them as their own kittens are removed after weaning. This doesn't seem to do the mother any harm, provided, of course, she gets enough to eat.

Both mother cat and kittens are unusually trusting during the first few weeks after birth. For the mother, this is due to a wave of the hormone oxytocin, which drives her to make her kittens her top priority. For her kittens, the factor may not be a hormone such as oxytocin, but instead their early inability to produce stress hormones such as adrenalin. A kitten that suckles for a few seconds too long and is accidentally dragged out of the nest as its mother leaves should be terrified; after all, this is potentially the most traumatic experience it has had in its short life. We can see the latent danger in this situation: the last thing such a kitten should do is make an association between the mother's smell and the shock of falling to the ground outside the nest. If it does, when the mother returns it might shy away from her, rather than immediately attaching to her to feed as it should. However, thankfully, kittens' inability to produce stress hormones means that such incidents leave no lasting impression.

Once a kitten reaches two weeks of age, its eyes and ears open and it begins to take its first faltering steps around the nest (see the figure on page 85). Its stress mechanisms then begin to function, en-

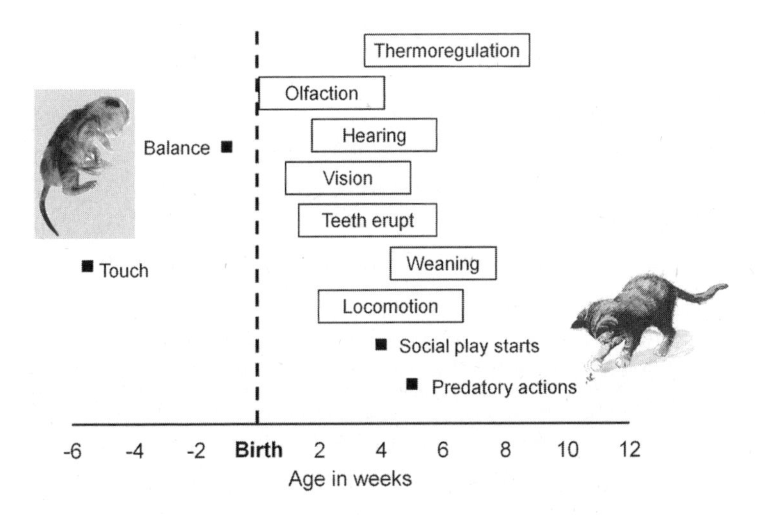

The development of the senses and other critical events in the life of a kitten

abling it to learn both what's bad and what's good about the world. From this point, what the kittens learn depends on whether their mother is present. If she's there, the kittens take their lead from her, but do not become stressed, so may not remember much about anything untoward that happens to them. Only if she is absent, and they must make their own decisions on how to react, do their stress levels rise abruptly, cementing their memory of the trauma and whether they dealt with it effectively or not.

This "social buffering" is probably adaptive for cats in the wild. A mother that is absent for long periods is probably having difficulty finding enough food for her litter or to replenish her milk, so the kittens must start to learn about the world soon if they are to stand any chance of survival. Kittens whose mother is usually with them in the nest can delay learning about any perils that may await them, because they can rely on their mother to protect them.

The personalities of cats are powerfully affected by what they learn when they're tiny kittens. Most kittens born in houses will be looked after both by their mothers and by their mothers' owners, but those unfortunate enough to undergo prolonged stress in their first few weeks of life may grow up to have enduring emotional and cognitive problems. For example, kittens that are abandoned by their mothers and are then hand-raised can become excessively attention-seeking toward their first owners, though some subsequently seem to grow out of this. Based on what we know about other mammals in similar situations, we can assume that after the mother's departure, the kittens' brains endure high levels of stress hormones. These consistently high levels cause permanent changes in their developing brains and stress hormone systems, such that they may overreact to unsettling events later in life.

Such cats may not make particularly satisfying pets, but this is by no means to say that they are mentally defective. Rather, their apparently abnormal behavior is an evolved adaptation. A mother that has struggled to raise her kittens was likely affected by some difficulty that has made food hard to find. A kitten whose mother has struggled to raise it therefore expects to emerge into an uncertain world, in which it may have to live on its wits and thereby outcompete its littermates and any other kittens born in the vicinity. The kittens of a relaxed, well-fed mother are likely to be able to depend on a more stable world, one in which they will have the time to hone their social skills and reproduce several times over a period of years. Such kittens, of course, are likely to make better pets than kittens that have been stressed in early life.

As a kitten begins its third week, it embarks on the most crucial six weeks of its life, as far as its development is concerned (see box on page 87, "Stages of Development"). From this point onward, its eyes, ears, and legs begin to function reliably, and, guided by its hormones, it begins to make decisions about whom and what it should interact with, and whom and what it should keep away from. At the same time, its brain is growing rapidly, every day adding thousands of new nerve cells and millions of new connections among them, establishing the framework for storing all the new knowledge it accumulates.

Stages of Development

The way a cat reacts to the world around it develops for at least the first year of its life, but most of the crucial changes take place in the first three or four months. Biologists divide this into four periods, each of which has a different significance for the growing kitten.

During the *prenatal period*, especially the second month of the queen's pregnancy, the kitten is largely—but not entirely—isolated from the outside world. The composition of the amniotic fluid and the blood in the placenta both reflect the mother's environment. For example, if the mother eats a strongly flavored food during this period, the kittens may prefer to be weaned onto a food with the same flavor, showing that they gain the ability to learn well before they emerge into the world. Female kittens that are adjacent to male kittens in the womb absorb some of their testosterone and are briefly more aggressive in their social play than kittens in all-female litters. Such inclinations are likely to be short-lived, but more far-reaching changes are also possible. Based on what we know of other mammals, if the mother is highly stressed during her pregnancy, her stress hormones may cross the placenta and impair the development of her kittens' brains and endocrine systems.

During most of the *neonatal period*, from birth to about two and a half weeks of age, the kitten is deaf and blind, relying on its senses of smell and touch to bond to its mother.

The *socialization period* begins as the eyes and ears open and begin to function in the third week of life, enabling the kitten to start learning about the world around it, including the people that are caring for it and its mother. At the same time, it learns to walk and then to run. When they are not sleeping, kittens spend much of this period engaged in play, initially with one another and then increasingly with objects.

The beginning of the *juvenile period*, at eight weeks of age, coincides with the customary time to rehome a kitten (except pedigree kittens, traditionally homed at thirteen weeks). By this point, the sensitive period for socialization is virtually over. The juvenile period ends at sexual maturity, sometime between seven months and one year of age; many pet cats will of course be neutered before the end of the first year.

Its mother is still a crucial influence during this period, but from now on a kitten gradually becomes able to distinguish its littermates from one another, and to learn about the other animals around it, including people. Kittens born in the wild also begin to learn how to hunt, for within a few short weeks they will need to feed themselves.

Most of the kittens' interactions with one another are playful, and for the first half of the socialization period, most of their play is directed at other kittens. However, we do not know if early on each kitten recognizes that what it is playing with is another kitten; most of the actions performed are similar to those used later toward objects. Bouts of play are short and disorganized, and it may be that the movements of the other kitten trigger each attempt at play. By the time they are six weeks old, though, kittens will play on their own with the objects around them, poking, pouncing, chasing, batting them with their paws, and tossing them in the air. These are all actions adult cats use when they capture prey, so biologists have looked for an elusive link between the amount of play that kittens perform and their hunting ability when they grow up. While play with objects hones the kittens' general coordination, it's probably not the most important factor in determining whether a cat grows up to be capable of catching enough prey to keep itself fed.

For feral kittens, their mother enables them to learn how to fend for themselves. As soon as they are old enough, she brings recently killed prey back to the nest; as they become better coordinated, she brings back prey that is still alive. This gives the kittens the opportunity both to handle prey and to find out what it tastes like. The mother doesn't seem actively to teach them how to deal with prey; rather, she simply places it in front of them and allows their predatory instincts to take over. If they show no interest, she may draw attention to the prey by starting to feed on it herself, stopping when the kittens begin to join in. Of course, this process rarely happens with owned cats, unless the mother happens to be an accomplished and habitual hunter herself, in which case small, gory "presents" may find their way into the nest.

Whether a cat is destined to be a hunter or (more likely) not, among the most important events in a kitten's life is when its mother decides to start weaning it. This typically occurs in the fourth or fifth week of the kitten's life, but may be earlier if the litter is large—six or more kittens—or if the mother is unwell or stressed. Whatever the circumstances, the mother cat drives this process; kittens rarely if ever decide to wean themselves. At the decisive moment, the mother starts to spend time away from the kittens or simply blocks access to her milk by lying or crouching with her abdomen pressed firmly to the ground. Not surprisingly, the kittens begin to get hungry, and for a few days their weight gain, steady since birth, slows down or even stops. Hunger drives them to become much more inquisitive about other possible sources of food.

In the home, with human owners supplying the grub, the new source of food should be special kitten food. In the wild, mother cats bring prey back to the nest and dissect it, making it easier for small mouths to chew. The kittens continue to pester her for milk, but for the next couple of weeks or so she will ration them, to force them to develop their ability to eat—and digest—meat. As their eating habits change, so do their insides. Meat takes longer to digest than milk, so kittens' intestines become lined with villi, small finger-shaped projections that increase the amount of nutrients that can be absorbed. Lactase, the enzyme that breaks down milk sugar, is permanently replaced by sucrase to break down the sugars in muscle, such that many adult cats find milk indigestible. The mother cat is simply being cruel to be kind: when the kittens become fully weaned at about eight weeks of age, she may spontaneously allow them to suckle occasionally, possibly as a way of reinforcing family ties—although in a domestic situation, they may no longer be with her.

Scientists have sometimes portrayed the weaning process as a conflict between mother and offspring. One theory holds that an animal such as a cat that can have several litters in a lifetime should behave in a way that balances out the survival of each of her litters. For example, the demands of an extra-large litter might become excessive, jeopardizing her own health and therefore her chance of having any more litters. In some mammals, including mice and prairie dogs, mothers with large litters may kill one or two of the weakest members, presumably to ensure the survival of the rest. However, this tactic has never been recorded with cats, although a sickly kitten may simply be ignored by its mother, presumably because it does not produce the right signals to induce her to care for it.

Each kitten must do its best to survive, regardless of whether it is its mother's favorite; if it does not survive to maturity, it will never leave any offspring of its own. Given that female cats tend toward promiscuity, no kitten can be sure that it is closely related even to its littermates, let alone to any of the members of its mother's next litter.[8] It therefore has to put its own interests first, even more so than if it could be sure that it shared both parents with its littermates. It therefore has no incentive to give up suckling, even to the point of weakening its mother.

A mother cat cannot afford to be too hardhearted toward her offspring; after all, she cannot be at all certain that she will have the chance to breed again. She therefore carefully assesses their needs, keeping them hungry enough to want to try meat while not significantly compromising their health. For example, if her milk temporarily dries up before she's finished weaning them, she will start nursing them again before weaning is over, as soon as her milk comes in again; this ensures that that they don't miss out on any essential nutrients. Likewise, the kittens themselves cannot be too aggressive in demanding milk from their mother, for (in the wild) they need to keep her on their side for several weeks longer, as she teaches them essential hunting skills. Furthermore, if her milk supply starts to fail early, kittens *increase* the amount of time they spend playing, presumably to prepare themselves for learning how to hunt. While most young animals—mice, for example—play *less* when they're hungry, presumably to conserve energy, for kittens play is preparation for hunting behavior. Responding to their mother's predicament, such kittens are thereby preparing themselves for a premature independence.

Play prepares kittens not just for hunting, but also for getting along with other cats. If domestic cats were as solitary as their ancestors, they would have little need for social graces. For animals whose individual contact is restricted to brief courtship and mating, and then the raising of each litter by its mother, social play is unsophisticated and brief. For domestic kittens, play with littermates becomes increasingly sophisticated as they get older, and no longer revolves around the elements of hunting behavior.

By about six to eight weeks of age, kittens begin to use specific signals aimed at persuading their littermates to play with them, such as rolling onto their backs (see "Belly-Up" in the drawing on page 92), placing their mouth over another kitten's neck ("Stand-Up"), or rearing onto their hind legs ("Vertical Stance"). By ten weeks—assuming the litter is still together, since many kittens are homed at eight weeks old—each kitten will have come to learn the "correct" response to each of these—Belly-Up for Stand-Up (and vice versa), and Belly-Up for Vertical Stance. As the kittens get older, play tends to get rougher, and occasionally one of the kittens gets hurt. To avoid any confusion between play and a real fight, kittens will use a "Play Face" to indicate their friendly intentions, particularly when in the vulnerable Belly-Up position. They may also use special movements of their tails to signal playfulness, but so far no scientist has been able to decode these. Older kittens also have a special signal showing when they want to stop playing: they arch their back, curl their tail upward, and then leap off the ground.

If a litter is left together after the normal age for homing, social play occupies more and more of the kittens' time, peaking somewhere between nine and fourteen weeks of age. All this sophistication confirms that domestic kittens are designed to become social adults, a process that begins when they are just a few weeks old and, unless interrupted, continues for several months.

Surprisingly, we know little about the optimum time for cats to learn how to interact with other cats. Experiments pinpointing the sensitive period for socialization to people have yet to be repeated to

Kittens performing Belly-Up with a Play-Face (left) and Stand-Up (right)

uncover the same for cat interaction, but we can assume that there is more than one sensitive period, each tailored to the social environment in which the kitten finds itself. The first period spans the kitten's first two weeks, when kittens form their attachment to their mother, based on olfaction.

During the first four weeks or so of life, each kitten learns how to interact with its siblings; it may have little need to recognize its littermates as individuals, but that probably follows soon after. Kittens are likely born with a template as to what another kitten looks like, but this is easily overwritten if there are no other kittens available. Thus, a kitten raised in a litter of puppies accepts the puppies as its littermates, and does not appear to "know" that it itself is a kitten. However, if a puppy is introduced into a litter of kittens, even though they are perfectly friendly toward the puppy, the kittens still prefer one another's company. Cats' brains must be constructed in such a way that they form stronger attachments to cats than to other four-legged animals.

From the fifth week onward, kittens certainly learn a great deal from their littermates, in particular the most effective way to play. If a kitten from a litter of one is introduced to a kitten that has grown up with other kittens, it will play much more roughly than normal. Hand-raised kittens are even more inept: some turn out to be so aggressive that other kittens actively avoid them. Others become excessively bonded to their human owners and seem barely to realize that they are cats at all.[9] We do not yet know why some go one way and some the other, but possibly some important interactions between socialization to cats and socialization to people help to produce a cat that doesn't overreact to new situations—a balanced individual, if you like. Hand-reared kittens may develop extreme personalities because they miss out on these interactions, due to their lack of contact with other cats.

Kitten inviting play using Vertical Stance

Littermates that are homed together usually form a stronger bond with one another than two unrelated cats. In August and September of 1998, a student and I studied this by recording the behavior of pairs of cats in boarding catteries (people who have a pair of cats that get along together at home will usually ask for them to be housed together when they board them). We compared fourteen pairs of littermates that had lived together since birth with eleven pairs of unrelated individuals that had not met each other until at least one of the pair was more than one year old. Despite the hot weather, all the littermate pairs slept in contact with each other, but we observed only five of the unrelated pairs ever lying in contact with each other, and even those only occasionally. Many of the littermate pairs groomed each other; the unrelated pairs never did. Almost all the littermates were happy to feed side by side; we had to feed most of the unrelated pairs from separate bowls or in turns.[10]

This study does not clarify whether just being littermates makes cats friendly to one another, but this seems the most likely explanation. It's unlikely that it was simply the age difference between the unrelated pairs that made them unfriendly, since if a kitten is kept with its mother rather than being homed, the mother and kitten generally remain friends for life. However, scientists have yet to investigate whether littermates can recognize one another as relations, say if they are separated and reintroduced months later (as dogs can), or whether cats use a simple rule of paw, and try to remain friends with any cat that they've lived with continuously since the second month of their life—that is, during the socialization period.

Unlike most kittens born in homes, feral kittens can continue to interact with their littermates, and any other nearby litters, until they are at least six months old. Most kittens born outdoors are born in the spring. By autumn, their mother severs all ties with her male offspring and may actively drive them away—a sensible precaution against risk of inbreeding. Up to that point, the kittens will have had many opportunities to learn more about what it means to be a cat, opportunities generally denied to pet cats of the same age. Female kittens, on the other hand, often do not leave their natal group until they are several years old, so they have even more scope for learning feline social graces.

We often see a corresponding difference between the sexes in kittens' choice of company. As they grow, feral male kittens spend most of their time with brothers from the same litter. They rarely interact with kittens from other litters, even those they are related to; in any one year, most litters born within a feral colony will be first or second cousins. Male cats, assuming they avoid being neutered, are destined to lead solitary lives, and when they reach maturity, must compete with one another for female attention. Female feral kittens initially spend most time with their littermate sisters, but at a few months old will probably also regularly interact with both their aunts and their aunts' or other female relatives' kittens.

The third and fourth months of a kitten's life are full of play, regardless of whether they are male or female. We do not know whether denying kittens the opportunity to play with other kittens during this period has significant consequences. Perhaps because cats are often conceived of as solitary creatures for which a social life is a luxury, not a necessity, we have not scientifically investigated this topic. However, it seems possible that continued interaction with their peer group during adolescence could make a major contribution to cats' development as social animals.

The most important social skill a cat must learn to become a pet is, of course, not how to interact with cats (though that is useful), but how to interact with people. In this respect, cats rank second only to dogs. Like dogs, cats can learn how to behave toward their own kind and toward people not only virtually simultaneously, but also without confusing the two. Nearly all other domestic animals are not as adaptable. For example, a lamb that has to be hand-raised will attach itself very powerfully to the person that feeds it, as if that person was its mother. Unless introduced to other, mother-reared lambs as quickly as possible, its behavior may be abnormal for the rest of its life. Moreover, whether sheep are tame or not, their social behavior is always primarily directed at other sheep.

Cats, like dogs, are capable of multiple socialization, the ability to become attached to animals of several different species—and not just people and other cats. Kittens raised in a household with a cat-friendly dog will continue to be friendly toward that dog, and potentially other similar dogs, for their entire lives. We do not know precisely how cats (or dogs) achieve this, but we can speculate that cats keep the "rules" for interacting with each separate species in discrete parts of their brains, as the human brain stores different languages in physically distinct areas of the frontal lobes.

Between the ages of four and eight weeks, kittens form their view of people, or at least the people they meet. Kittens that meet only women during that period, as can happen in some breeding catteries, may turn out fearful of men or children once they are homed. A kitten that is handled by only one person may become very strongly attached to that person, purring whenever it is picked up and in-

Kittens with a dog

sistently pestering that person for attention—an intensity that suggests that the person has, in the kitten's mind, taken the place of its mother.[11]

Introducing a kitten to a wide variety of people before the age of eight weeks seems to produce an approachable cat. Doing so seems to block the development of a strong attachment to one person, and instead builds a general picture of the human race in the cat's mind. Whether (say) three categories—men, women, children—develop simultaneously, or whether kittens learn to place all humans in a single category is unknown, but the end result is self-evident: a cat that is not fearful of humans.

Kittens need a lot of daily exposure to people to become optimally socialized to them. In one study, fifteen minutes handling each day produced a kitten that would approach people, but not as enthusiastically as a kitten that had been handled for forty minutes per day. Likewise, the fifteen-minute kitten would not stay on a lap for as long as the forty-minute kitten.[12] Fortunately for them, most kittens born in homes get this much attention without any special effort being taken, thanks to their irresistible cuteness.

Kittens born in animal shelters may not have the same luxury. Concerns over transmission of disease between litters may stand in the way of optimal socialization. The standard way of looking after mother-cats with kittens, in which most contact with people occurs during routine feeding and cleaning, produces reasonably friendly kittens, but these will almost inevitably not have as much human contact as kittens born in a home. However, veterinary behavior specialist Rachel Casey and I found that additional handling and play, involving several people rather than just one—even as little as a few minutes extra each day—from their third week to their ninth week, produced a dramatic improvement in how friendly these kittens were. This increased interaction then affected their relationships with the people who adopted them, who were kept unaware as to whether their kitten had received extra handling. As they reached their first birthdays, the kittens given extra handling were noticeably more relaxed than those raised in the standard way; likewise, their owners reported that they felt closer to them. The extra handling produced a long-term effect on the strength of the bond between cat and owner.[13]

In the UK, eight weeks old—the end of weaning—is the traditional time to home a random-bred cat, but neither our studies, nor any other published research has yet addressed whether eight weeks of age is the right time for a kitten to move to its new home—especially from the viewpoint of how well it will bond with its new owners. From the new owner's perspective, this age seems entirely understandable: after all, kittens are at their cutest at around eight weeks of age. However, pedigree cats are generally not homed until they are thirteen weeks of age; the Governing Council of the Cat Fancy, one of the bodies that regulates cat breeding in the UK, strongly recommends that no kitten should be permitted to go to a new home any earlier than this, because until then it has not received its initial course of vaccinations and is therefore susceptible to disease. Unfortunately, this difference in traditions cannot tell us whether age of homing affects the relationship with the new owner. The personalities of the main types of pedigree breeds—Persians/longhairs and Orientals/Siamese—differ so much from one another and from most house cats that it would swamp any impact of even a four-week difference in the age of homing.

Kittens of any age must surely be affected by being moved to a new home. Everything they have come to know disappears at a stroke, and everything about their new home is novel. They leave the security of their mother, who has probably recently finished weaning them and has just relaxed the arm's-length attitude she had used to persuade them to accept solid food. They are plucked from the company of their littermates, whom they have gotten to know as individuals over the past month, and who have been their willing play partners. They leave the security of surroundings that they know well and that smells reassuringly of their mother, brothers, and sisters. The attention of unfamiliar humans, however well-meaning, is unlikely to comfort them during this period of change.[14]

Whether such dislocation happens at eight or thirteen weeks, or sometime in between, it takes place when the kitten is much younger than the age at which it would spontaneously leave its family group.

It's a tribute to the cat's behavioral flexibility that this process works at all; indeed, provided that they receive a reasonable amount of handling from their mothers' owners, most kittens evidently adapt to their new surroundings and become attached to their new owners. They may also—although this is by no means certain—adapt to living alongside their owners' other cats.

Handling between four and eight weeks of age seems essential to a kitten becoming a contented pet. But what happens when handling is delayed until six, seven, or eight weeks, or even later? Many kittens are born to mothers who, because they are feral or stray, are wary of people, and are not discovered until they begin to make their own first faltering steps into the world. In the 1990s, I worked with the UK charity Cats Protection to study this topic. Rehoming charities are frequently called in when such kittens are found, and naturally these organizations wish to help. What is the best course of action?

Our study found that the older the kitten when it was first handled, the less friendly it seemed to be—at least to begin with. Kittens that had received no human contact until they were six weeks old behaved distinctly from normal kittens, even after they had settled into their new surroundings at the rehoming centers. If rescued at six weeks, they were not easy to handle, and very few purred when stroked.[15] Kittens not rescued until eight weeks of age were difficult to handle, and those not found until ten weeks were, at least to begin with, virtually wild. The exceptions were a few litters of kittens that, although they had not been picked up for the first time until they were eleven weeks old, had allowed themselves to be stroked occasionally while in their nests a few weeks earlier. These kittens behaved more like those rescued at seven weeks—initially wary of people and difficult to handle. This confirms that socialization to people has to *start* within the first six or seven weeks if it is to become effective, but that once started the process can continue for several weeks more, especially if initial exposure has been brief.

The way the kittens were handled once rescued also affected how quickly they became friendly. If they had been handled by two or more people, they were more relaxed and playful when introduced to unfamiliar people, than if they had been looked after by only one person. Again, it appears that attachment to an individual and socialization to people in general progress in parallel at this age. Kittens who know only one person become attached to that particular person, but may remain wary of other humans; kittens who meet several different people at roughly the same time may not become so attached, but are later much more accepting of people in general.

Most of these rescued feral kittens became perfectly satisfactory pets, the same as kittens who were born in the same rescue centers and were therefore handled from a very early age. In fact, the feral kittens had received *extra* attention to compensate for the socialization they hadn't been able to receive before they were rescued, so many turned out to be more friendly at one year old than their shelter-born counterparts. However, the few kittens that were difficult to socialize at all were still unapproachable at the same age.[16]

Kittens that don't meet a human until the age of ten weeks or older are unlikely to become pets, except in extreme circumstances. Instead, they live as "stray" or "feral" cats, living on the fringes of human activity but never becoming part of it. Most hunt to some extent, but virtually all depend on food and shelter provided by people, both accidentally and deliberately. Their only window of opportunity for accepting and then becoming attached to people has passed. The cat's social brain changes suddenly at about eight weeks of age, and altering its basic social inclinations after that is usually impossible.

The general rule is, once a feral, always a feral—unless it experiences severe physical and mental trauma, such as being hit by a motor vehicle. Occasionally, some kindhearted soul will take a feral cat that has been the victim of an accident to a veterinarian. Many such cats are beyond saving, but those that cheat death and are nursed slowly back to health can go through an unexpected change in personality; they become attached to whoever has looked after them the most, in a manner reminiscent of a hand-raised kitten. Researchers have recorded similar changes in cats that have suffered severe and protracted fevers. Apparently, the deluge of stress hormones released in a cat that is close to death can scramble the brain enough to go through the socialization process all over again.

The importance of the socialization period to a kitten's future welfare cannot be underestimated. In just six short weeks, beginning as it turns two weeks of age, this period constructs the foundations for all its subsequent social life. If the kitten is unlucky enough to have no brothers and sisters, and has no other kittens nearby, its view of what it is to be a cat is incomplete; although cat mothers will play with single kittens, they are much less inclined to do so than other kittens. If a kitten's mother keeps it away from people, usually because she is herself not socialized to humans, the kitten is unlikely to become a pet. If it is handled by only one person, it will become attached to that person but may form too narrow a view of what humans are like, and thus become wary of strangers. If the kitten has to be hand-reared, it misses out on learning how to be a cat and its entire social and cognitive development may be impaired.

Cats don't suddenly stop learning about people or other animals when they pass the eight-week watershed. At this point, the general course of their attachments is set, but the details of their journey through life are far from predetermined. We know that they learn a great deal more about how to interact with people throughout the first year of their life, and there's every reason to suppose that the same applies to how they adapt when they come into contact with other cats, although little research has been done on this. The way a kitten's personality develops over that first year depends not only on their experiences, but also on genetics: like other animals, individual cats adopt different strategies for dealing with the same events, and these differences are often the product of their genes.

CHAPTER 5

The World According to Cat

We may easily overlook the simple fact that cats live in a world that is—subjectively speaking—quite distinct from ours. We share just enough overlap between our different perceptions of the world to be able to relate to one another, each species having evolved its senses to fit its lifestyle. It is unreasonable to consider cats' abilities as inferior—or superior—to our own. Biologists abandoned the idea of one species being "superior" to another decades ago, though cat owners may suspect that their cats feel otherwise. We now consider each species as having evolved to fit a particular way of life; cats are good at being cats, because their ancestors evolved sense organs and brains to suit that role.

Because the domestication of the cat is still incomplete, that role is still in a state of flux—for example, cats are still adapting to urban living—while their sense organs have remained more or less unchanged. One major difference between cats and people is that cats have evolved genetically, from wild to domestic, while over the same timescale we ourselves have evolved culturally, from hunter-gatherer to city dweller. Genetic evolution is a much slower process than cultural evolution, and the 4,000 years over which cats have adapted to living alongside humankind is not long enough for any major change in sensory or mental abilities. Thus, cats today have essentially the same senses, the same brains, and the same emotional repertoire as their wildcat forebears: not enough time has passed for them to break free of their origins as hunters. As far as we know, all that has changed in their brains is a new ability to form social attachments to people, while their senses remain completely unaltered.

Some cat behavior seems baffling to us, but it may stem from their ability to sense things around them to which we are oblivious—and vice versa. A complete understanding of cats requires that we try to visualize the world they live in, which is a world quite different from what our own instincts tell us it should be. Indeed, I use the word "visualize" because that's how our imaginations work: we conjure pictures in our heads of past events, or of what might happen in the future. Scientists doubt that cats' brains work that way; not only is it unlikely that their brains are capable of such "time travel," but also their world, unlike ours, is not based on appearance. Smell is at least as important to cats as vision is, so even if they could imagine, they might well conjure what something smells like rather than what it looks like. A few humans can do this—professional perfumers and sommeliers, for example—but usually only after extensive training.

This fundamental emphasis on other senses is not the only difference in the way cats and humans perceive the world. Each of our individual senses also works differently too, so that, for example, a cat and a person looking out of the same window will see two dissimilar pictures.

Human eyes and cat eyes share some similarities—we're both mammals, after all—but cat eyes have evolved as superefficient aids to hunting prey. The wild ancestors of the modern cat needed to maximize the time they could spend hunting, so their eyes enabled them to see in the merest glimmer of light. This has affected the structure of cats' eyes in several ways. First, they are huge in comparison with the size of their heads; indeed, their eyes are almost the same actual size as our own. In darkness, their pupils expand to three times the area ours do. The efficiency with which their eyes capture light is further enhanced by a reflective layer behind the retina, known as the tapetum. Any incoming light that misses the receptor cells in the retina bounces off the tapetum and back through the retina again, where some will happen to strike a receptor cell from behind, enhancing the sensitivity of the eye by up to 40 percent. Any light that misses the second time around will pass back out through the pupil, giving the cat its characteristic green eye shine whenever a light is shone into its eyes in the dark.

The receptor cells on the retina are also arranged differently from ours. They fall into the same two basic types—rods for black-and-white vision in dim light, and cones for color vision when it's bright—but cats' eyes have mainly rods, whereas ours have mainly cones. Instead of each rod connecting to a single nerve, cats' rods are first connected together in bundles; as a result, cats' eyes have ten times fewer nerves traveling between their eyes and their brains than ours. The advantage of this arrangement is that the cat can see in the near-dark, when our eyes are nearly useless. The disadvantage is that in brighter light, cats miss out on finer details; their brains are not being told precisely which rod is firing, only a general area on the retina onto which the light is falling.

The result of this disadvantage is that in full daylight, cats cannot see as well as we can. The rods become overloaded, as ours do under the same conditions, and have to be switched off. The small number of cones that cats do have are spread all over their retinas, rather than concentrated in the center of the retina, in the fovea, as ours are, so they get a general and not very detailed picture of their surroundings during the day. Because their pupils are so large when wide open, they cannot be shrunk to a pinprick in bright sunlight, as ours can. Instead, cats have evolved the ability to contract their pupils to narrow vertical slits, less than 1⁄32 of an inch wide, which protects their sensitive retinas from being overwhelmed with light. They can further reduce the amount of light entering by half-shutting their eyes, thereby covering the top and bottom of the slit and leaving only the center exposed.

Cats also show little interest in color; among mammals, color seems a uniquely primate, especially human, obsession.[1] Like dogs, cats have only two types of cones and see only two colors, blue and yellow; in humans, we call this red-green color blindness. To cats, both red and green probably look grayish.[2] Moreover, even colors they can distinguish seem to be of little relevance to them. Their brains contain only a few nerves dedicated to color comparisons, and it is difficult to train cats to distinguish between blue and yellow objects. Any other difference between objects—brightness, pattern, shape, or size—seems to matter more to cats than does color.

Another drawback of having such large eyes is that they are not easy to focus. We have muscles in our eyes that distort the shape of the lens to allow close vision; cats seem to have to move their whole lens back and forth, as happens in a camera, a much more cumbersome process. Perhaps because it is just too much effort, they often don't bother to focus at all, unless something exciting, such as a bird flying past, catches their attention. Close focus, anything nearer than about a foot away, is also out of the question with such large eyes. Furthermore, the muscles that focus the lens seem to set themselves according to the environment the cat grows up in: outdoor cats are slightly longsighted, whereas all-indoor cats tend to be shortsighted. Despite the largeness of their eyes, cats can swivel them quickly to keep track of rapidly moving prey. To avoid image blurring, the eyes do not move smoothly but in a series of jerks, known as saccades, about a quarter of a second apart, so that the cat's brain can process each separate image clearly.

Like humans, cats have binocular vision. The signals from each of their forward-facing eyes are matched up in their brains and are converted there into three-dimensional pictures. Most mammalian carnivores have eyes that point forward to provide them with binocular vision, so that they can judge precisely how far away potential prey is, and judge their pounce accordingly. Presumably because their eyes don't focus any closer than about a foot away from their noses, cats don't bother to converge their eyes on objects any closer than this.[3] To compensate, cats can swing their whiskers forward to provide a 3-D tactile "picture" of objects that are right in front of their noses.

Binocular vision is the best way of judging how far away something is, but it's not the only method available. Cats that lose an eye because of disease or injury can compensate by making exaggerated bobbing movements of their heads, monitoring how the images of the various objects they can see move relative to one another. Prey animals such as rabbits commonly do this: because their eyes are on the sides of their heads to maximize surveillance, they have little or no binocular vision, and have to rely on other, slightly cruder ways of judging distance.

The cat's ability to detect tiny movements is another legacy of its predatory past. The visual cortex, the part of the brain that receives signals from the eyes, does not simply construct pictures as if the eyes were two still cameras; it also analyzes what has changed between one picture and the next. The cat's visual cortex compares these "pictures" sixty times each second—slightly more frequently than our visual cortex does, meaning that cats see fluorescent lights and older TV screens as flickery. Dedicated brain cells analyze movements in various directions—up and down, left to right, and along both diagonals—and even local brightening or dimming of specific parts of the image. Thus, the most important features of the image—the parts that are changing rapidly—are instantly singled out for attention.

Cats learn how to integrate all this information when they are kittens—unlike amphibians, for example, which already have specialized prey-detector circuits formed in their brains when they metamorphose from tadpole to adult. Cats use their movement detectors to behave flexibly when they're hunting, paying equal attention to a mouse it spots attempting to make an escape, or to a movement of the grass that betrays the mouse's position. Both help the hunting cat to find a meal.

We can plainly see the cat's origins as a predator of small rodents in its remarkable hearing abilities—remarkable both in the range of sounds it can hear and in pinpointing the source of the sound. The cat's hearing range extends two octaves higher than ours, into the region that—because we can't hear it—we refer to as ultrasound. This extended range enables cats to hear the ultrasonic pulses bats use to orient themselves while flying in the dark, and the high-pitched squeaks of mice and other small rodents. Cats can also tell different types of rodents apart by their squeaks.

In addition to this sensitivity to ultrasound, cats can hear the same full range of frequencies we can, from the lowest bass notes to the highest treble. Almost no other mammal exhibits such a wide range, about eleven octaves in total. Because cats' heads are smaller than ours, their hearing range should be shifted to higher frequencies, so their ability to hear ultrasound is perhaps not all that remarkable; rather, it's their ability to hear low notes that is unexpected. The cat's ability to hear sounds lower than it should, based on the size of its head, is possible because they have an exceptionally large resonating chamber behind the eardrum. The capacity to hear ultrasound despite this arises from a feature of this chamber not seen in other mammals: it divides into two interconnecting compartments, thereby increasing the range of frequencies over which the eardrum will vibrate.

Mobile, erect ears are the cat's direction finders, essential when tracking a mouse rustling through the undergrowth. Cats' brains analyze the differences between the sounds reaching the right and left ear, enabling the cat to pinpoint the source. For lower-pitched sounds that fall into our hearing range—for example, when we talk to our cats—the sound arrives at one ear slightly out of sync with the other. Also, higher frequencies are muffled by the time they reach the ear farthest from the source, providing a further clue to where the source is. This is essentially the way that we too determine where a sound is coming from, but cats have an extra trick: the external parts of the ears are independently mobile, and can be pointed at or away from the sound to confirm its direction. When it comes to ultrasounds, that are above our hearing range, such as a mouse's squeak, the phase differences become too small to be useful, but the muffling effect gets larger and therefore becomes more informative. Therefore, a cat has little difficulty determining whether a sound is coming from the right or the left.

In addition, the structure of their external ears—the visible part of the ears, technically referred to as pinnae—also enables cats to tell with some accuracy how high up the source of a sound is. First and foremost, the corrugations inside the pinnae add stiffness and keep the ears upright, but they also cause complex changes to any sound as it passes into the ear canal; these changes vary depending on how far above or below the cat the sound is coming from. Somehow, the cat's brain decodes these changes, which must be difficult, given that the pinnae may be moving. The pinnae are also directional amplifiers, but rather than being tuned to pick up mouse squeaks, they are especially sensitive to the frequencies found in other cats' vocalizations, enabling male cats to pick up the calls made by females as they come into season, and vice versa. This is perhaps the only feature of the cat's ears not refined specifically for detecting prey.

Cats' hearing is therefore superior to ours in many ways, but inferior in one respect: the ability to distinguish minor differences between sounds, both in pitch and intensity. If it was possible to train a cat to sing, it couldn't sing in tune (bad news for Andrew Lloyd Webber). Human ears are outstanding at telling similar sounds apart, probably an adaptation to our use of speech to communicate, and, within that, our ability to recognize subtle intricacies of intonation that indicate the emotional content of what we are hearing—even when the speaker is trying to disguise his or her voice. Such subtleties are probably lost on cats, although they do seem to prefer us to talk to them in a high-pitched voice. Perhaps gruff male voices remind them of the rumbling growl of an angry tomcat.

As with hearing, the cat's sense of touch features refinements that help with hunting. Cats' paws are exceptionally sensitive, which explains why many cats don't like having their feet handled. Not only are a cat's pads packed with receptors that tell it what is beneath or between its paws, but the claws are also packed with nerve endings that enable the cat to know both how far each claw has been extended and how much resistance it is experiencing. Since wild cats generally first catch their prey with their forepaws before biting, their pads and claws must provide essential clues on the efforts the prey is making to escape. Cats' long canine teeth are also especially sensitive to touch, enabling the hunting cat to direct its killing bite accurately, sliding one of these teeth between the vertebrae on its victim's neck and killing it instantly and almost painlessly. The bite itself is triggered by special receptors on the snout and the lips, which tell the cat precisely when to open and then close its mouth.

The cat's whiskers are basically modified hairs, but where the whiskers attach to the skin around the muzzle they are equipped with receptors that tell the cat how far each whisker is being bent back, and how quickly. Although cat's whiskers are not as mobile as a rat's, a cat can sweep its whiskers both forward, compensating for longsightedness when pouncing, and backward, to prevent the whiskers from being damaged in a fight. Cats also have tufts of stiffened hairs just above the eyes, triggering the blink reflex if the eyes are threatened, and on the sides of the head and near the ankles. All of these, in tandem with the whiskers, enable cats to judge the width of openings they can squeeze through.

Information gathered from these hairs help keep the cat upright, but the vestibular system, in the inner ear, contributes most to the cat's exquisite sense of balance. Unlike our other senses, balance operates almost entirely at the subconscious level, and we barely notice it until something causes it to malfunction—for example, motion sickness. Although the information that the cat's vestibular system produces is used more effectively, it is actually similar to ours.

This system consists of five fluid-filled tubes. In each, sensory hairs on the inside detect any movement of the fluid, which occurs only when the cat's head twists suddenly; because of inertia, the fluid doesn't move as quickly as the sides of the tube do, dragging the hairs to one side (if you're reading this with a cup of coffee in front of you, try gently rotating the cup: the liquid in the middle of the cup remains where it is). Three of the tubes are curved into half-circles, aligned at right angles to one another to detect movement in all three dimensions. In the other two, the hairs are attached to tiny crystals, which make the hairs hang downward under gravity, enabling the cat to know which way is up, and also how fast it is moving forward.

One reason cats are agile is simply that they walk on four legs rather than two. Four legs need coordinating if they are to work effectively as a team, and the cat has two separate groups of nerves that do this. One group relays information about each leg's position to the other three, without involving the brain; the other sends information to the brain for comparison with what the inner-ear balance organ relates about the cat's position. More reflexes in the neck enable the cat to hold its head steady even when it is moving quickly over uneven ground—a necessity for keeping its eyes on its prey.

When walking from place to place, cats pay close attention to where they are going. Because of their poor close vision, they have little reason to look down at their front feet, so instead they look three or four paces ahead and briefly memorize the terrain in front, allowing them to step over any obstacles in their path. Scientists have recently determined that if a cat is distracted with a dish of tasty food while walking, it forgets what the ground in its path looks like and has to have another look before setting off again. In the experiment, researchers switched off overhead lights while the cat was distracted into looking to one side; it then had to feel its way gingerly forward, indicating that its view of the path had vanished from its short-term memory. However if the cat was distracted after it had stepped over an obstacle with its front paws, and while the obstacle was right under its belly, it remembered that it should lift its hind paws when it started walking again, even after a ten-minute delay—and even if the obstacle, unbeknown to the cat, had been moved out of the way. Somehow the visual memory of the obstacle is converted from ephemeral to long-lasting by the simple act of stepping over it with the front feet.[4]

A cat's gravity-detecting system is most impressive when it either jumps voluntarily or accidentally slips and falls. Less than a tenth of a second after all four feet lose contact with a surface, the balance organs sense which way up the head is, and reflexes cause the neck to rotate so that the cat can then look downward toward where it will land. Other reflexes cause first the forelegs and then hind legs to rotate to point downward. All this happens in thin air, with nothing for the cat to push on. While the front legs are being rotated, they tuck up to reduce their inertia, while the back legs remain extended; then the front legs are extended while the back legs are briefly tucked up (see the nearby figure). Ice skaters use the same principle to speed up spins, simply by retracting their arms and the spare leg. The cat also briefly curves its flexible back away from its feet as it rotates, which helps to prevent the twist at the back end cancelling out the twist at the front.[5] Many cats also counter-rotate their tails to stabilize their fall. Finally, all four legs extend in preparation for landing, while the back is arched to cushion the impact.

How a cat rights itself after a sudden fall

While this intricate midair ballet is happening, the cat could have already fallen as far as ten feet. As such, it's possible for a short fall to injure the cat as much as, and possibly more than, a longer fall, if there is insufficient time for the cat to prepare itself for landing. If a cat falls out of a high-rise building or a tall tree, it has another trick available: forming a "parachute" by spreading all four legs out sideways, before adopting the landing position at the last minute. Laboratory simulations suggest that this limits the falling speed to a maximum of fifty-three miles per hour. This tactic apparently allows some cats to survive falls from high buildings with only minor injuries.

Like dogs, cats rely greatly on their sense of smell. Cats' balance, hearing, and night vision are all superior to our own, but it's in their sense of smell that they really outperform humans. Everybody knows that dogs have excellent noses, something humankind has made use of for millennia, and this prowess is located, in part, in their large olfactory bulbs, the part of the brain where smells are first analyzed. Relative to their size, cats have smaller olfactory bulbs than dogs, but theirs are still considerably larger than ours. Although scientists have not studied the cat's olfactory ability in as much detail as the dog's, we have no reason to suppose that a cat's sense of smell is much less acute. Without a doubt, it is certainly much better than our own.

Like those of dogs, the insides of cats' noses have far more surface area devoted to trapping smells than ours do—about five times as much. Indeed, it's *Homo sapiens* who appears particularly deficient in this respect. During the evolution of our primate ancestors, we seem to have traded most of our olfactory ability for the benefits of three-color vision, which evolutionists theorize enabled us to discriminate red ripe fruits and tender pink leaves from their generally less-nutritious green counterparts. Cats' sense of smell is more or less typical of mammals, and dogs' is more acute than the average. As in most mammals, air passing into the nose is first cleaned, moistened, and warmed if necessary as it passes over skin supported on a delicate honeycomb of bones, the maxilloturbinals. The air then reaches the surface that extracts and decodes the odor, the olfactory membrane, which is supported on another bony maze, the ethmoturbinals. Because, unlike dogs, cats do not pursue their prey over long distances, their maxilloturbinals are not especially large; dogs must sniff and run at the same time, and while they're doing this their olfactory membranes are constantly at risk from damage by dust, or dry or cold air. Cats' habit of sitting and waiting for their prey places much less strain on the air-conditioning system in their noses.

Nerve endings in the olfactory membrane trap the molecules that make up the smell. The tips of the nerves are far too delicate to come into contact with air themselves, so they are covered in a protective film of mucus, through which the molecules pass. This film has to be very thin; otherwise, the molecules would take several seconds to move from the airflow to the nerve endings. If this was the case, then the information conveyed by the odor would be out of date before the cat knew it was there. To facilitate a speedy response, the mucus has to be spread so thinly that the nerve endings become damaged from time to time—for example, they may dry out when they become temporarily exposed to the air—and they therefore regenerate about once a month.

The other ends of the olfactory nerves are connected together in bundles of between 10 and 100 before transmitting their information to the brain. Cats have several hundred kinds of olfactory receptors, and the information arises from whichever of these has been triggered by the odor passing through the nose. Each bundle contains only nerves with the same kind of receptor, to amplify the signal without muddling up the data it contains. In the brain, the input from the different receptors is compared to build up a picture of the odor in question.

This system is unlike that of the eye, where images are built up by each section of the retina transmitting its information directly to the brain. The nose does not build up a "two-dimensional" image, as each eye does; as the cat breathes in, the air is swirled around so much in the nostrils that whatever receptor each odor molecule strikes is a matter of pure chance. It's even unclear whether, unlike their vision and hearing, cats can make any sense of the slightly different amounts of odor entering their left and right nostrils.

Cats are probably capable of distinguishing among many thousands of different smells, so they cannot have one receptor dedicated to each one. Rather, cats deduce the character of each odor they encounter from which type of receptor is being stimulated, and by how much, in comparison with other types. Although scientists do not yet know precisely how the resulting information is combined together in any species of mammal, the potential resolution of such a system is staggering. Consider that our brains can generate a million or so distinct colors from just three types of cones. Several hundred olfactory receptors must therefore have the capacity to discriminate among billions of different odors. Whether cats achieve this is difficult to say;

we do not even know precisely how many different odors can be discriminated by our own noses, and we have only about one-third to a half the number of receptor types cats have. Based on these extrapolations, the mammalian olfactory receptor system seems rather overengineered, and science has not yet resolved why this might be. Suffice it to say, that cats should theoretically be able to distinguish between more smells than it is likely to encounter in a lifetime.

Scientists know little about how cats make use of their sensitive noses. Cats' most dramatic response is to the odor of catnip, but this seems to be an aberration (see box on page 115, "Catnip and Other Stimulants"). We know a great deal about the olfactory capabilities of dogs because we have harnessed dogs' noses for various purposes: finding game, tracking fugitives, and detecting contraband, to name but three. If cats were as easy to train as dogs are, we'd probably discover that their olfactory performance is close to that of dogs. A few minutes cursory observation of any cat will reveal that it sniffs its surroundings all the time, confirming that it places a premium on what things smell like. Remarkably, however, it was only in 2010 that the first scientific account of cats using olfaction in hunting was published.[6]

This study showed that cats do indeed locate prey using their scent marks. Many of the rodents that cats hunt, especially mice, communicate with one another using scent signals carried in their urine. As mammals, the noses of cats and mice work in much the same way, so it is highly unlikely that mice can disguise their scent marks so that cats cannot detect them. Australian biologists proved this by collecting sand from mouse cages and placing it on the ground on roadside shoulders. Almost all these patches of sand were visited by predators—mostly foxes, but tracks of feral cats were also apparent—while clean sand was not. The collected data did not show how far away the cats had traveled from, but it is possible that significant distances were involved—that is, the cats probably navigated upwind toward the odor sources, rather than only investigating the patches of sand because they looked unusual. We know that many dogs prefer to hunt using their noses, and can detect and then locate sources of odor from hundreds of feet away. While cats prefer

Catnip and Other Stimulants

Scientists do not yet understand why cats respond to catnip, a traditional constituent of cat toys. Not all cats respond to it. A single gene governs whether or not the cat responds, and in many cats, perhaps as many as one in three, both copies of this gene are defective, with no apparent effect on behavior or general health.

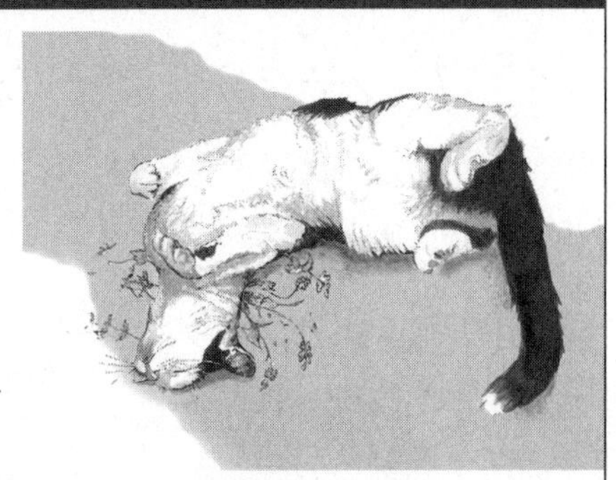

Rolling on catnip

The behavior released by catnip is a bizarre mixture of play, feeding, and female sexual behavior, whether the cat itself is male or female. Cats may first play with a catnip toy as if they think it is a small item of prey, but they quickly switch into bouts of a seemingly ecstatic combination of face-rubbing and body-rolling, reminiscent of a female cat in season. Most cats also drool and attempt to lick the catnip. This behavior may continue for several minutes at a time, until the cat eventually recovers and walks away—but if the toy is left where it is, the cat may repeat the whole sequence, albeit with less intensity, twenty or thirty minutes later.

A few other plants elicit the same response, notably the Japanese cat shrub or silver vine, and the roots of the kiwifruit vine, which despite its name originated in southern China. In the 1970s, the first growers of kiwi vines in France learned this when they found to their distress that cats had excavated and chewed their seedlings. All three plants contain similar fragrant chemicals, thought to be responsible for the cats' responses.

By some accident of evolution, these chemicals likely stimulate the cat's nose to trigger circuits in the brain that would never normally be activated at the same time, somehow bypassing the normal mechanisms that ensure that cats don't perform two incompatible actions at once. A cat in the throes of catnip-induced oblivion would seem vulnerable to attack, and since cats presumably don't get any lasting benefit from their experience, evolution should have weeded out the gene responsible. Most species in the cat family, from lions to domestic cats, respond to these plants the same way, so the gene must have evolved several million years ago. Why it did so remains a mystery.

to use vision when hunting in daylight, they probably switch to using their sense of smell when hunting at night, when sight, even their sensitive night vision, becomes less reliable.

Finding prey from the odors it produces can be difficult. Scent marks rarely indicate the current position of the animal that left them, only where it was when it made the mark, perhaps hours earlier. In the sand experiment, the urine samples continued to attract cats for at least a couple of days. It's possible that cats, as sit-and-wait hunters, use the scent marks to see whether they will attract other members of the same species that made them. Mice use urine marks to signal to other mice, and the marks contain a great deal of useful information about the mouse that produced them. Thus, the scent-marker doesn't put itself at risk, so much as other members of the same species that show up to examine the mark.

In addition, odors spread out from their sources in ways that are not predictable. We know intuitively that light travels in straight lines but not around obstacles, and that sound travels in all directions, including around obstacles. However, because we rely so little on odor to give us directional information, it's not immediately obvious what problems animals face when determining where an odor is emanating from. Of course, odors outdoors are carried by the air—downwind, not upwind—but air movements close to the ground, where cats operate, are usually highly complex. While the wind may be blowing in a consistent direction a few yards above the ground, friction caused by its contact with the ground and especially with vegetation causes it to break up into eddies of various sizes. These carry "pockets" of odor away from the source, so that a cat somewhere downwind of a mouse's nest will get intermittent bursts of mouse smell.

Tracing these bursts to their source, especially in thick cover, is likely to require diligent searching and possibly some backtracking. Once a cat has located a source of mouse odor, it is potentially able to use the fact that odors don't travel upwind to position itself downwind of the odor source so its own smell can't be detected by the mouse, and then wait for mice to turn up. While sit-and-wait is a well-documented feline hunting method, we do not know whether cats routinely prefer patrolling the downwind sides of, for example, hedgerows to avoid their own scent betraying their presence. However, it seems likely that a predator as smart as a cat could quickly learn this tactic, even if it is not instinctive.

Cats possess a second olfactory apparatus, which humans lack: the vomeronasal organ (VNO; also called the Jacobson's organ).[7] A pair of tubes, the nasopalatine canals, run from the roof of the cat's mouth, just behind the upper incisors, up to the nostrils; connected roughly halfway up each one of these tubes is a sac, the VNO itself, filled with chemical receptors. Unlike the nose, the entire VNO is full of fluid, so odors must be dissolved in saliva before they can be detected. Moreover, the ducts connecting the VNOs to the canals are only about one-hundredth of an inch wide, so thin that the odors must be pumped in and out of the sacs by a dedicated set of tiny muscles.[8] This gives the cat precise control over when it uses the VNO—unlike the nose, which automatically receives odor every time the cat breathes. Thus, the VNO's function lies somewhere between our senses of smell and taste. Appreciating how cats make use of this faculty requires a great leap of imagination.

Cats, unlike dogs, perform an obvious facial contortion when bringing the VNO into play. They pull their top lip upward slightly, uncovering the top teeth, while the mouth is held partially open. This pose is usually held for several seconds: it's sometimes known as the "gape" response, although it's usually referred to by its German name, "Flehmen." Researchers theorize that during this pose, the tongue is squeezing saliva up into the canals, from where the pumping mechanism delivers it to the VNO.

Cats perform Flehmen exclusively in social situations, so by implication they must use their VNO to detect the smells of other cats.[9] Male cats perform it after sniffing urine marks left by females, including during courtship, and female cats will do the same toward urine marks left by tomcats, although usually only if the tom is not present.

The cat's VNO can probably detect and analyze a wide range of "smells," since it contains at least thirty different kinds of receptors; more than the dog's, which has only nine. These receptors are distinct

Cat displaying "Flehmen"—using its vomeronasal organ to detect the odor left by a cat that has cheek-rubbed the twig

from those found in the nose, and are connected to their own dedicated area of the brain, known as the accessory olfactory bulb.

Why do cats—and, indeed, most mammals, apart from primates—need two olfactory systems? The answer seems to change from species to species. Mice have highly sophisticated VNOs, with several hundred receptor types and two distinct connections to the accessory olfactory bulb, rather than the cat's one; the odorants mice pick up regulate reproduction, as well as enabling recognition of every other mouse in the neighborhood from its unique odor "fingerprint." In many species, some odor communication takes place through the VNO and some through the nose. For example, in rabbits, chemical communication between adults involves the VNO, but the scent emanating from the mother that stimulates her kits to suckle is picked up by their noses. Sometimes the balance between the two changes as the animal matures: the VNO and the nose are used in tandem during a guinea pig's first breeding season, but the following year its nose alone will suffice. Although cats have not been studied in this amount of detail, it's feasible that they also interpret olfactory information in a flexible way.

If we accept that the VNO is primarily designed to analyze odors coming from other members of the same species, then the fact that dogs are generally *more* sociable than cats doesn't really fit the fact that their VNOs are *less* discriminating. Dogs, descended from social ancestors, conduct much of their relationships with other dogs face to face, and thus use visual cues to confirm who another dog is and what it is likely to do next, so their VNOs may not be required very often.

The domestic cat's solitary ancestors only rarely had the opportunity to meet one another; the exceptions are males when they are courting females, and females interacting with their litters for a few months. In the wild, much of the social life of cats must be conducted through scent marks, which can be deposited for another cat to sniff days, sometimes weeks later. Since wild cats rarely get to meet other members of their own species, any information they can pick up from scent marks is crucial for making decisions on how to act when they do encounter each other. Most critically for the survival of her offspring, a female cat must assess her various male suitors, who themselves have been attracted by the change in her own scent as she comes into season. She may already have gained useful information about each of them by sniffing the scent marks they have left as they roamed through her territory, information she can use to supplement what she can see of their condition and behavior when she finally meets them. She may also be able to distinguish those that are unrelated to her from those that are—perhaps a son who has roamed away and then happened to return to the area a few years later—thereby avoiding inbreeding. Scientists have not yet studied any of these possibilities in cats, but we know them to occur in other species.

Cats' sense of smell has evolved not purely for hunting, but also for social purposes. Successful hunting was, until domestication, crucial to the survival of every individual cat. However, it cannot in itself ensure the survival of an individual cat's genes; that also requires an effective mating strategy. Each female cat tries to select the best male for her purposes every time she mates to ensure that her genes pass down to successive generations. Ideally, she should take a long-term view, trying to gauge not only how likely her own offspring are to survive, but also how successful they are likely to be when they in turn become old enough to breed. If she picks the strongest and healthiest male(s) to mate with, then her male kittens are likely to be strong and healthy also when they are old enough to mate. She will of course be able to make a judgment based on her suitors' appearance, but she may be able to obtain a better idea of how healthy they are based on their odor. Her sense of smell may thus provide her with extra information to make these crucial mating decisions.

Cats probably perceive most of the scents that have social meaning using the nose and the VNO in tandem. While both may be needed the first time a particular smell is encountered—for example, the first time a young male detects a female in estrus—on subsequent occasions either one will do the job, the brain presumably using memories of previous encounters to "fill in" the missing input.

Like dogs, cats pay a great deal of attention to scent marks left by other cats, both those carried in urine and those that they rub onto prominent objects, using the glands around the mouth. To distinguish it from the rubbing that cats perform on people and on one another, which is mainly a tactile display, this is sometimes referred to as "bunting," a term whose origins are obscure. Cats' faces have numerous scent-producing glands, one under the chin, one at each corner of the mouth, and one beneath the areas of sparse fur between eye and ear, while the pinna itself produces a characteristic odor. We know little about how cats use these scent marks, but they certainly display an interest in the scents other cats produce. For example, male cats can distinguish between females at different stages in their estrus cycle based on their facial gland secretions alone. Each gland produces a unique blend of chemicals, some of which have even been used in commercial products that can have a positive effect on reducing stress in anxious cats.[10]

Apart from the social role of the VNO, and to a lesser extent the nose, all the cat's other senses are exquisitely tuned to the hunting lifestyle of their ancestors. They have quite an arsenal at their disposal: they can locate prey visually, their eyes effective in the half-light of early dawn and late dusk; aurally, detecting high-pitched squeaks and rustles; or olfactorily, through detecting the odors that rodents leave in scent marks. As they approach their prey, cats' exquisite sense of balance and the sensory hairs on their cheeks and elbows allow them to do so silently and stealthily. As they pounce, the whiskers on their faces sweep forward to act as a short-range radar, guiding the mouth and teeth to precisely the right place to deliver the killing bite. Cats evolved as hunters, something domestication has done little to change.

What the cat senses is only half the story. Their brains have to make sense of the vast amounts of information that their eyes, ears, balance organs, noses, and whiskers produce, and then turn that information into action, whether correcting the cat's balance as it tiptoes along the top of a fence, deciding on the precise moment to pounce on a mouse, or checking the yard for the scent of cats that have visited during the night. The sheer volume of data that each sense organ generates has to be filtered every waking second. An analogy might be the vast banks of TV screens and monitors at NASA headquarters during the launch of a spacecraft: at any one moment, only a minute fraction of what they display is important, and it takes a highly trained observer to know which ones to watch and which can safely be ignored. Unfortunately at present we know much less about how sensory information is processed than how it is generated.

The size and organization of the cat's brain can give us some clues as to their priorities in life. The basic form of the felid brain, as shown by the shape of the skull, evolved at least 5 million years ago. Some parts of the brain, especially the cerebellum, are disproportionally geared to processing information relating to balance and movement, reflecting cats' prowess as athletes. While this is apparently contradicted by the occasions that cats get stuck up trees, the problem here is not their intelligence or their sense of balance, but rather that their claws all face forward, so they cannot be used as brakes when descending. The part of the cortex that deals with hearing is well-developed; so too, as we have seen, are the olfactory bulbs.

In cats, the parts of the brain that seem to be important in regulating social interactions are also less well-developed than they are in the most social members of the Carnivora, such as the wolf and the African hunting dog. This is unsurprising, given the solitary lifestyle of the domestic cat's immediate ancestors. Nevertheless, domestic cats are remarkably adaptable in their social arrangements; some form deep attachments with people, and others remain in a colony with other cats for their entire lives, their only interactions with humans consisting of running away and hiding. Once made, these choices cannot be reversed, since they are set during socialization: the cat as a species can adapt to a number of social environments, but individual cats generally cannot. This lack of flexibility must ultimately lie in the way that their brains are constructed, and in particular, the parts of their brains that process social information. Science has yet to unravel the factors that lie behind these constraints, so that today's cats have limited options when faced with changes in their social milieu.

CHAPTER 6

Thoughts and Feelings

Historically, scientists have avoided words such as "thinking" and "feeling" when talking about animals. "Thinking" runs the risk of being too imprecise: it can mean anything from simply paying attention to something ("I'm thinking about cats"), to complex comparisons between memories and projections into the future ("I'm thinking about the best way to get my cat to come in at night"), to expressions of opinion ("I think that cats are such fussy eaters because they have such unusual nutritional needs"). To avoid the implication that animals such as cats possess human-type consciousness, biologists tend to use the term "cognition" to refer to their mental processing of information.

With "feelings," our intuitive grasp of our own emotions is bound up in our consciousness: we are aware of our emotions to an extent that cats almost certainly are not.[1] However, new scientific techniques such as brain imaging have revealed that all mammals, and therefore cats, have the mental machinery necessary to produce many of the same emotions we feel, even though they probably experience them in a much more in-the-moment way than we do. We do not have to presume that cats are conscious animals to allow that they are capable of making decisions—decisions based not just on information they are receiving and their memories of similar events, but also their emotional reactions to that information. In other words, it's now scientifically acceptable to explain their behavior in terms of what they "think" and "feel" as long as we bear in mind that cats' thought processes and their emotional lives are both significantly different from our own.

Bearing this in mind is a challenge: we are accustomed to thinking about cat behavior on our own terms. Part of the pleasure of owning a pet comes from projecting our thoughts and feelings onto the animal, treating it as if it were almost human. We talk to our cats as if they could understand our every word, while knowing full well that they certainly can't. We use adjectives like "aloof" and "mischievous" and "sly" to describe cats—well, other people's cats, anyway—without really knowing whether these are just how we imagine the cat to be, or whether the cat knows it possesses these qualities (and is secretly proud of them).

Nearly a century ago, pioneering psychologist Leonard Trelawny Hobhouse wrote, "I once had a cat which learned to 'knock at the door' by lifting the mat outside and letting it fall. The common account of this proceeding would be that the cat did it to get in. It assumes the cat's action to be determined by its end. Is the common account wrong?"[2] As this illustrates, scientists have long struggled to find a coherent way to interpret cats' behavior rationally and objectively. Scientists still argue the extent to which cats and other mammals can solve problems by thinking them through in advance, as we do. We can easily interpret cat behavior as if it had purpose behind it, but is this mere anthropomorphism? Are we assuming that because we would solve a problem in a particular way, cats must be using similar mental processes? Often, we find that cats can solve what appear to be difficult problems by applying much simpler learning processes.

Cognitive processes—"thoughts"—begin in the sense organs and end in memory. At every stage, information is filtered out: there is simply not enough room in the cat's brain (or for that matter the human brain) to store a representation of every scrap of data picked up by its sense organs. Some of the filtering takes place as the sense organs relay their information to the brain; for example, the motion detectors in the cat's visual cortex draw attention to what is changing in the cat's field of vision, enabling it for an instant to ignore everything else. Within the brain, representations of what is happening are generated and held for a few seconds in *working memory* before most are discarded. A small fraction of these representations, particularly those that have triggered changes in emotion, transfer into *long-term memory*, enabling them to be recalled later on. Short-term memory, long-term memory, and emotion are all used when a cat needs to make a decision as to what action to take.

Much of the everyday behavior we observe in our pet cats can be explained by simple mental processes. First, the information gathered by the sense organs has to be categorized: is the animal over there a rat, or could it be a mouse? Then, it must be compared with the situation as it was a few moments before: Has the rat moved, or is it still in the same place? More or less at the same time, the cat's long-term memory is being trawled for similar situations: What happened last time it saw a rat?

As far as we can tell, recollection of such memories affects the cat's decisions through two mechanisms. The first is an emotional reaction: a cat that has been bitten by rats in the past will immediately feel fear and/or anxiety, and a cat skilled at killing and eating rats will feel something like excitement. The second mechanism guides the cat in selecting the most appropriate action for the situation—depending on the emotional reaction, either the best way of getting out of the rat's way or the hunting tactic that has worked best on previous rats.

Our minds continually categorize objects without being aware of what we are doing, a process that requires sophisticated mental processes. Scientists are now studying whether cats' minds use the same processes, and if their brains can fill in gaps, as ours can. Let's say a cat sees a mouse's nose and tail, but the mouse's body is obscured behind a plant. Can the cat imagine the mouse's body in between, or does it perceive the nose and tail as somehow belonging to two separate animals? Cats can be trained to distinguish drawings that—to our eyes—create visual illusions, such as the one shown, from those that do not, so it is likely they can indeed "join the dots," and visualize the body of the mouse between its head and tail. Cats can also use changes in texture to piece together shapes of particular interest to them—to a cat, a negative image of a bird, with the contrasts shifted the wrong way around, is still recognizably a bird.[3] However, they do not seem to have an inbuilt rat/mouse detector as such—unlike toads, for example, which reflexively pounce on anything wormlike.

Cats presumably do not know in advance what type of prey is likely to be available when they first leave their mothers and start hunting for themselves, so they rely on what they've learned as kittens, rather than robotically pouncing on mice or other prey.

Cats can also make sophisticated judgments on how large or small something is. If they are trained to pick out the smallest or largest of three objects, they continue to pick out the smallest when all three objects are made smaller, so that what was originally the small object is no longer the smallest of the three. Prey will appear larger or smaller depending on how far away it is, so making a judgment about relative size is important in deciding whether to run away (from a large rat, still some distance away) or attack (a small rat, close by). Mysteriously, cats also seem to classify shapes according to whether

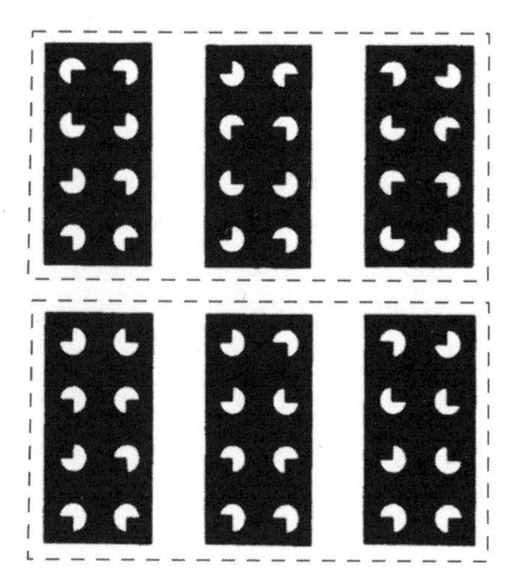

Cats can recognize outlines even when they are broken up or unusual. They know the difference between pictures that produce an illusion, like the "falling square" in the three pictures at top left, and those that don't, like the three at bottom left. They also recognize the negative image as a bird.

they are closed—for example, a filled circle or square—or open—for example, an uppercase I or U. We do not know why this skill evolved as it did, since its contribution to cats' survival is obscure.

That all these examples relate to vision is a consequence of our own biases: because we are a visual species, scientists tend to focus on an animal's visual abilities to get an idea of how their brains work. Cats must also be able to classify what they hear, and although we don't know how they categorize sounds, we can guess from their hunting behavior that they probably have categories for each of the sounds made by the various species they prey upon. Presumably they also have categories for the odors that they pick up with their noses and their vomeronasal organs, but with our comparatively poor sense of smell, we have trouble imagining how such a system might work.

Humans also categorize events by when they happened, but cats probably do not. We know little about cats' conception of time, but they are definitely much better at judging short durations than long ones. Cats have been successfully trained to discriminate sounds that last four seconds from those that last five, and also to delay their response to a cue for a few seconds (because they only get the reward if they wait for the correct time).[4] However, cats are poor at discriminating longer periods of time, and their perception is likely limited to the few seconds provided by their working memory. We have no evidence to suggest that cats can spontaneously recall memories and place those events as having happened a few days ago, as opposed to a few hours or weeks previous—something we find easy to do.

Cats have a general sense of the rhythm of the day. They have a free-running daily rhythm that is reset every day by the onset of daylight, and they also take other cues from their environment about what time of day it is. Some are natural, such as the sun rising and setting, and some are learned, such as their owner feeding them at roughly the same time every day. Still, they don't seem to think *about* time passing, in the way that we do.

Once a cat has worked out what it is observing through sight, smell, or hearing, it must work out what to do next. If its survival might be threatened, the cat may need to act first and think later. When a cat is startled, say by a sudden loud noise, it instantly prepares itself for action through a set of preprogrammed and coordinated reflexes. It crouches, ready to run if necessary. Its pupils dilate while its eyes quickly focus as closely as possible, regardless of whether there is anything there to focus on; this presumably maximizes the chances of pinpointing the threat if it is close by. If the threat is still far away, then the cat has less urgency to identify what it is.

Almost every other reaction a cat makes changes with experience: over time, its reactions change. Even the startle reflex gradually wanes and may eventually disappear, say if the same loud noise is repeated over and over again. In this process, habituation, something initially excites the cat, but then becomes progressively less interesting until it eventually evokes no reaction at all.

For example, cats are renowned for quickly getting "bored" with toys. Intrigued as to why this should be, in 1992 I set up a research project at the University of Southampton to look into cats' motivation for playing with objects. Do they literally "play" for the sheer fun of it, as a child might, or are their intentions more "serious"? The manner in which cats play with toys is highly reminiscent of the way that they attack prey, so we designed our experiments with the presumption that whatever was going on in their heads, it was probably related to their hunting instincts. My graduate student Sarah Hall and I found that habituation is the main underlying reason for this apparent boredom. We presented cats with toys—mouse-sized, fake-fur-covered "pillows" tied to a piece of cord—and at first they usually played intensely, appearing to treat the toy as if it was indeed a mouse. However, many cats stopped playing within a matter of a couple minutes. When we took the toys away for a while and then presented them again, most of the cats started playing again, but neither as intensely nor for as long as the first time. By the third presentation, many of the cats would scarcely even begin to play. They clearly became "bored" with the toy.

If we switched the toy for a slightly different one—a different color (say, black to white, since cats' perception of colors is different from ours), texture, or odor—almost all of the cats would start playing again. Thus, they were "bored" not by the game, but by the toy itself. In fact, the frustration of being offered the same toy repeatedly actually *increased* their desire to play. If the interval between the last game with the original toy and the first game with the new toy was about five minutes, they attacked the second toy with even more vigor than they did the first one.[5]

To understand why playing with a toy would make a cat frustrated, we considered what might motivate cats to play in the first place. Kittens sometimes play with toys as if they were fellow kittens, but adult cats invariably treat toys as if they were prey: they chase, bite, claw, and pounce on toys just as if the toys were mice or rats. To test the idea that cats think of toys in the same way they think of prey, we tried different kinds of toys to see which ones cats prefer. Our findings showed that, unsurprisingly, they like mouse-sized toys that are furry, feathered, or multi-legged—toy spiders, for example. Even indoor cats that had never hunted showed these preferences, so they must be hardwired in the cat's brain. The cats played with rat-sized toys covered in fake fur in a subtly different way from the mouse-sized toys. Instead of holding them in their front paws and biting them, most cats would hold the rat-sized toys at arm's length and rake them with their hind claws—just as hunting cats do with real rats. The cats were apparently thinking of their toys as if they were real animals, and as if their size, texture, and any simulated movement (such as our pulling on the toy's string) had triggered hunting instincts.

We then examined whether a cat's appetite has similar effects on the way it hunts and the way it plays with toys. If cats play with toys just for their own amusement, as many people assume they do, then they should be less inclined to play when they are hungry, since their minds should be focused instead on how to get something to eat. Conversely, as a hunting cat gets hungrier, it will hunt more intensely and become more inclined to take on larger prey than usual. We found exactly the latter when we offered toys to our cats. If their first meal of the day had been delayed, they played more intensely than usual with a mouse-sized toy—for example, biting it more frequently.

Moreover, many of the cats that normally refused to play with a rat-sized toy at all were now prepared to attack it.[6] This convinced us that adult cats do think that they are hunting when they're playing with toys.

Cats don't easily get "bored" with hunting, so we were still puzzled as to why our cats stopped playing with most toys so quickly. Indeed, they appeared to get "bored" with most commercially available toys and with the kinds of toys we made for our first experiments. The few toys that sustained our cats' interest all shared one quality: they fell apart as the cat was playing with them.[7] Although we had to abandon experiments that involved these toys, which came apart at the seams as our cats batted them about, we noticed that several of the cats were extremely reluctant to give them up. We then realized that our original swapping experiments mimicked one aspect of what happens when a cat rips a toy apart: when we exchanged the toy for a slightly different one, the cat's senses told it that the toy had changed. It didn't seem to matter to the cat that it had not caused the change itself; what was important was that a change seemed to have occurred.

We deduced that not only do cats think they are hunting when they're playing with toys, but their behavior is being controlled by the same four mechanisms whether they're hunting or playing. One of these mechanisms is affected by hunger, and the same one that makes a cat more likely to play with a toy makes it likely to make a kill when it's hungry.[8] The second is triggered by the appearance—and presumably the smell and sound—of prey, and certain specific features, such as fur, feathers, and legs, that the cat recognizes instinctively are likely to belong to prey animals. The third mechanism is affected by the size of the toy or prey. Attacking a mouse puts the cat in much less danger than attacking a rat, so the cat attacks the rat much more carefully; likewise, cats treat large toys much more circumspectly than small toys, as if they were capable of fighting back. Even though cats should quickly learn that the toys are unlikely to retaliate, most cats don't seem to do so. The fourth mechanism is the source of the cat's apparent frustration: if all that biting and clawing doesn't seem to have any effect on its target, then either the target wasn't a meal, or if it *is* prey, then it's proving difficult to subdue. A toy that starts to disintegrate, or is taken away but looks different when it comes back (as in our original experiment), mimics the early stages of a kill, thus encouraging the cat to persist.

Overall, many of the cat's hunting tactics can be explained in terms of simple reflexes, modified by emotion—specifically, the fear of getting hurt by large prey animals—and habituation, which ensures that the cat continues to grapple with its prey only if it is likely to end up with a meal. However, these are only the basic building blocks of hunting behavior; cats undoubtedly perfect their hunting skills through practice, by learning how to assemble the various elements in the most productive ways.

Habituation can explain many short-term changes in a cat's behavior, but its effects wear off after a few minutes; longer-term, more permanent changes in the way a cat reacts require a different explanation. These must be based on learning and memory.

Fundamentally, cats learn the same way as dogs, even though dogs are self-evidently much easier to train. Two factors lie behind this difference between cats and dogs. First, most cats do not find human attention rewarding in its own right, whereas dogs do; we therefore train cats using food as a reward, rather than affection. Second, dogs instinctively behave in ways we can easily shape into something useful: for example, the herding behavior of a sheepdog is composed of elements from the hunting behavior of the wolf, the dog's ancestor. Cat behavior features little that we can usefully refine by training, except for our own amusement. Obviously, we have benefited from the cat's hunting abilities for ages. However, we usually leave cats to their own devices: they will seek out the mice that invade our grain stores regardless of whether we want them to do so. Dogs, on the other hand, are specialized for cooperative hunting of much larger prey. They are a nuisance when unsupervised and useful only when trained. We take on the responsibility of gearing their attention toward particular prey, when that is what we need from them.

Much of what cats learn is based on two fundamental psychological processes: classical and operant conditioning. Both of these involve new associations forming in the cat's mind. The first involves two events that regularly occur closely together in time; the second involves something the cat does or does not do, and a predictable consequence of that action, which may be good for the cat (a reward), or bad (a punishment). Because cats seem to have little or no instinctive appreciation either of how humans behave or the best ways to interact with them, virtually all their dealings with us are built up through this sort of learning.

Classical conditioning is also known as Pavlovian conditioning, after Ivan Pavlov, the first scientist to map out how such learning works in a series of experiments with dogs in the 1890s. In fact, his principles apply equally well to cats.[9] A hungry cat that smells food will instinctively seek it out and then eat it. For a wild cat, food is the result of a successful hunting expedition; for a pet cat, the owner makes this trip unnecessary by buying food at the supermarket and presenting it to the cat. Cats don't need to learn that food can appear without being preceded by hunting, because this is precisely what happens when, in the wild, their mother brings food back to the nest. What they do learn, via classical conditioning, are the cues that indicate that food is on its way—for example, the sound of a can opener. In psychologist's jargon, this action by the owner is the conditioned stimulus, which becomes associated in the cat's mind with the unconditioned ("instinctive") stimulus, the smell of the food. Nothing in the cat's evolution has prepared it to respond automatically to the sound of a can opener: the association is something that every cat would have to learn for itself. Of course, this is hardly a difficult lesson, nor is the underlying process complex; scientists have found such behavior even in bees and caterpillars. Nevertheless, classical conditioning is the main way that cats find out how the world around them is constructed: which parts occur in predictable sequences, and which do not.

The result—the unconditioned response—doesn't have to be a "reward," such as food. In fact, learning occurs more quickly when it helps the animal to avoid something unpleasant or painful. A cat that is attacked by another, larger cat will certainly experience fear and possibly pain, and will instinctively try to run away. It will also probably remember what the attacking cat looked like, associating its appearance with the unpleasant feelings it experienced at the time.[10] The next time it sees that cat, it will feel the fear before any attack takes place—and may immediately run away, as it ultimately did the first time they met. However, relatively sophisticated animals such as cats can respond flexibly: they do not automatically have to perform the original response simply because stimuli are similar. In this way, the cat may not run away immediately, but instead "freeze," hoping to avoid detection, having previously learned that running away can invite a chase.

This simple learning has one major constraint: the events that a cat associates together must occur either at precisely the same moment or no more than a second or two apart. Say a cat has done something that its owner doesn't like—for example, depositing a dead mouse on the floor. The cat's owner finds the mouse several minutes after the cat has left it there and shouts at the cat. In this case, classical conditioning does not link the two events together: rather, it links the unpleasantness of being shouted at with whatever happened immediately before the shout—probably the owner's arrival in the room.

This rule has one exception: if a cat eats something that makes it feel ill, it will thereafter avoid foods with the same flavor. Moreover, forging this association requires only one such experience. This food-aversion learning differs from classical conditioning both in the speed with which the lesson is learned and in the delay between the sensation—the odor of the food—and the consequence—the upset stomach. It is obviously in the cat's best interest to avoid repeating any action that could kill it, which accounts for the irreversibility of this learning. Likewise, there would be little point in the cat associating its first feelings of sickness with something else that occurred at the same time—the problem food would have been eaten many minutes, even hours, before. Nevertheless, this is still classical conditioning, except that the "rule" for the time frame has been both extended and made much more specific, to the flavor of the last food that the cat ate before feeling ill; other cues, such as the characteristics of the room where it ate the meal, are ignored as irrelevant. Of course, feeling nauseous can also be a symptom of an infection unconnected with something the cat ate, so this mechanism occasionally has unexpected consequences: a cat that succumbs to a virus may then go off its regular food even after it has recovered, because it has incorrectly associated the illness with the meal that happened to precede it.

Cats can also learn spontaneously, when there is no obvious reward or penalty involved. This becomes especially useful when they are building up a mental map of their surroundings. A cat will learn that a particular shrub it passes every day has a particular smell. If the cat sees a similar shrub elsewhere, it will expect that shrub to have the same odor as the first one. If it turns out not to—perhaps because an unfamiliar animal has scent-marked it—the cat will give it an especially thorough inspection. Such "behaviorally silent" learning can be explained by classical Pavlovian learning—that is, if the cat spontaneously feels rewarded by the information it has gained. In other words, cats are programmed to enjoy their explorations; otherwise, they wouldn't learn anything from them.

This kind of learning allows cats to relax in what must be, for them, the highly artificial indoor environments we provide for them. Domesticated cats are happy once they have been able to set up a complete set of associations between what each feature of that environment looks, sounds, and smells like. This explains why cats immediately pay attention to anything that changes—move a piece of furniture from one side of the room to the other, and your cat, finding that its predictable set of associations have been broken, will feel compelled to inspect it carefully before it can settle down again. To cope with such changes, cats can gradually unlearn associations that no longer work, a process known technically as extinction.

All animals make some associations more easily than others; evolution has built in certain responses that are difficult to overwrite. For cats, one of these is the likelihood that high-pitched sounds are likely to come from prey. In one experiment, two Hungarian scientists taught some rats that whenever a high-pitched clicking sound came from a loudspeaker at one end of a corridor, a piece of food would appear at the other end, about seven feet away. The rats had to run quickly from the speaker to the food dispenser, otherwise the food would disappear. They learned this quickly and reliably, such that they would sit and wait for the clicking noise, and then on hearing it, run in the correct direction. To the experimenters' surprise, cats took a great deal longer to learn the same exercise: many of the cats required hundreds of repetitions before they would perform reliably. They quickly learned that food was available in the corridor when the clicking sound occurred, but almost invariably started running toward the sound rather than away from it in the direction of the food. While they were still learning the task, some of the cats appeared to become so conflicted that they refused to eat the food even if they reached it in time. Even when they had completely learned the task, they often glanced briefly at where the sound was coming from before proceeding toward the food, something the rats had given up doing at an early stage.[11]

The difference between the cats' and the rats' performance in this task does not mean that rats are smarter than cats. Rather, for a hungry cat, high-pitched noises are too important to ignore. For the rats, the clicking noise was an arbitrary cue, just as the sound of a can opener is for a cat; such a sound only comes to have meaning because of its association with the arrival of food. For the cats, the clicking noise instinctively meant "Food is likely to be over here," something they had great difficulty in learning to ignore.

Like other animals, cats can learn to perform a particular action every time a particular situation occurs. This forms the basis for much of animal training, and is technically known as operant conditioning. Contrary to popular myth, cats can be trained, but few people bother, apart from the professionals who produce performing cats for movies and TV. Cats are much more difficult to train than dogs are for at least three reasons. First, their behavior shows less intrinsic variety than that of dogs, so there is less raw material with which to work. Training any animal to do something that it would never do naturally is a difficult process; most training consists of changing the cues that lead to a piece of normal behavior, rather than inventing a piece of behavior the animal has never performed in its life. Second, and perhaps most important, cats are less naturally attentive toward people than dogs are. Domestic dogs have evolved to be exceptionally observant of what people want from them, because virtually every use that humankind has ever had for them has favored dogs that could interpret human behavior from those that could not. Cats are surprisingly good at following simple pointing gestures, but when they encounter a problem they cannot solve, they tend not to look to their owners for help—something that dogs automatically do.[12] Third, although dogs are powerfully rewarded by simple physical contact from their owner, few cats are: professional cat trainers generally have to rely on food rewards. These trainers also use secondary reinforcers extensively, rewards that are initially arbitrary but that become rewarding through association with the arrival of food. Nowadays, cat owners can emulate this using training aids such as clickers (see nearby box, "Clicker Training").

Cats can be trained to perform normal behavior out of its usual context. In the process known as shaping, the cat is initially rewarded for any behavior approximating the desired result that it performs immediately after the cue given by its trainer. Only behavior that is close to the result is rewarded, until finally only exactly the right response gets the reward. To take a simple example, most cats will (sensibly!) not jump over an obstacle if they can walk around it. To train a cat to jump on command, the trainer rewards it for walking over a stick that is lying on the ground, and then for stepping over it when it is raised slightly. Then if the trainer raises the stick further, she rewards the cat only for actual jumps. Once the habit is established, trainers can ingrain it further by rewarding only some successful performances and not others. This may seem counterintuitive, but animals usually concentrate harder when they know that the outcome is slightly uncertain than when it is guaranteed. Humans show this

Clicker Training

Like most animals (with the notable exception of dogs), cats can be trained only with food rewards. However, delivering the reward to the cat at exactly the right moment to reinforce the desired piece of behavior can be tricky; also, the cat may be distracted from what it is supposed to be learning by the smell of the food "concealed" in the trainer's hand. Cats can be trained much more easily if a secondary reinforcer is used—a distinctive cue that signals to the cat that a piece of food is on its way, and instantly making it feel good, thereby reinforcing the performance of whatever it was doing when the cue appeared.

Although in principle almost anything could be used as a secondary reinforcer, in practice distinctive sounds are the most convenient and practical, partly because they can be timed very precisely, and partly because the cat cannot avoid perceiving them even if it is some distance away and looking in the opposite direction. Animal trainers used to use whistles, but nowadays the cue of choice is often the clicker, a tensioned piece of metal in a plastic case that makes a distinctive click-clack sound when pressed and released.

Clicker-training a cat

Cats must be taught to like the sound of the clicker, which has little or no instinctive significance for them, by classical conditioning. This is simply done by attracting the cat's attention with a handful of its favorite

(continues)

Clicker Training *(continued)*

treats when it's hungry, and then offering the treats one at a time, each preceded with one click-clack from the clicker (Some cats are hypersensitive to metallic sounds—either hold the clicker away from the cat, or use a quieter sound, for example the plunger on a retractable ballpoint pen). After a few sessions, the sound of the clicker will have become firmly associated in the cat's mind with something pleasant, and this sound will gradually become pleasant in its own right.

Once this has become established, the clicker can be used to reward other pieces of behavior. For example, many cats can be trained to come consistently when called, initially by clicking when they turn and start to approach, but then gradually delaying the click until the cat arrives at the trainer's feet. Once the click has been established as a reward, it doesn't have to be followed by food every single time, although if the cat hears the sound over and over again without food appearing, the association may start to be lost. It is therefore usually most effective to intersperse sessions of training to come when called (when it is impossible to deliver the reward immediately after the click, because the cat is too far away) with repeat sessions of the original click-treat pairings that reestablish the link.

Instructions on how to clicker-train cats can be found at www.humanesociety.org/news/magazines/2011/05-06/it_all_clicks_together_join.html.

same quality under some circumstances—behavior that is perfectly exploited in the payout schedule of slot machines.

More complex tricks and "performances" are usually taught piece by piece, each step becoming linked together in the cat's mind through chaining. The easiest way to put a sequence together is to start at the end with the final action and its reward, and progressively add the preceding steps—backward chaining. For example, to train a cat to turn around once and then offer its paw to be shaken, the paw-shake is shaped first, and once that is perfected, the turn is shaped to precede it. Although it would seem more logical to train the first action first—forward chaining—most animals, including cats, find this much more difficult, indicating that their abilities to think ahead are limited.

Operant conditioning is not confined to deliberate training; it is one way that cats learn how to deal with whatever surroundings they find themselves in. Cats have not (yet) evolved to live in apartments; their instinctive behavior is still tuned to hunting in the open air. The fact that they can adapt to indoor living is testament to their learning abilities. Not only can they make sense of their surroundings by means of associations built up by classical conditioning, they are also able to learn how to manipulate objects around them to get what they want.

For example, many cats learn how to open a door fitted with a lever-type handle by jumping up and grabbing it with their forepaws. A superficially "clever" trick such as this one can be explained by operant conditioning. Of course, doors that unlatch with a handle and swing open on hinges do not form part of the world in which cats have done most of their evolving, so the final, successful, version of this behavior cannot be natural. However, it most likely starts with something that cats will do instinctively when they are unable to get somewhere they want to go, which is to jump up onto a vantage point to see whether there is an alternative route. If the cat tries to jump up onto the lever, which from the ground will look like a fixed platform, it will find that when the lever moves, not only does it lose its footing, but also that the door may swing open. The cat then gets the reward of being able to explore the room on the other side of the door—and cats, as territorial animals, find exploration of novel areas rewarding in itself. The cat will remember the association between its action and the reward. After trying various alternative actions on the handle, it will progressively arrive at the most efficient solution, which is to raise a single paw and gently pull the lever downward.

Pet cats learn how to use the same techniques on their owners. Even the most ardent cat-lovers sometimes describe them as manipulative, but much of a cat's so-called manipulative behavior is built up by operant conditioning. Feral cats are remarkably silent compared to domestic cats (except during fighting and courtship, notoriously noisy activities); in particular, such cats rarely meow at one another, whereas the meow is the pet cat's best-known call. The meow is usually directed at people, so rather than being an evolved signal it's more likely to have been shaped by some kind of reward.

Cats need to meow because we humans are generally so unobservant. Cats constantly monitor their surroundings (except when they're asleep, of course) but we often fix our gaze on newspapers and books, TVs and computer screens. We do, however, reliably look up when we hear something unusual, and cats quickly learn that a meow will grab our attention. For a few cats this may be rewarding in itself, but the meow will often also produce the reaction that the cat is hoping for, such as a bowl of food or an opened door. Some cats then shape their own behavior to increase the precision of their request. Some will deliver the meow at specific locations—by the door means "Let me out," and in the middle of the kitchen means "Feed me." Others find that different intonations lead to different results, and so "train" themselves to produce a whole range of different meows. These are generally different for every cat, and can be reliably interpreted only by the cat's owner, showing that each meow is an arbitrary, learned, attention-seeking sound rather than some universal cat-human "language."[13] Thus, a secret code of meows and other vocalizations develops between each cat and its owner, unique to that cat alone and meaning little to outsiders.

Classical and operant conditioning are not the only feasible explanations for why cats behave as they do, but they are often the simplest. Still, cats are undoubtedly much more than mere stimulus-response machines. Disentangling cat intelligence is a challenge, partially because we tend to believe willingly that much cat behavior is driven by rational thought. Even the earliest animal psychologists recognized this tendency. In his 1898 book *Animal Intelligence*, Edward L. Thorndike wrote, with his tongue firmly in his cheek:

> Thousands of cats on thousands of occasions sit helplessly yowling, and no one takes thought of it or writes to his friend, the professor; but let one cat claw at the knob of a door supposedly as a signal to be

> let out, and straightway this cat becomes the representative of the cat-mind in all the books.[14]

We also lack scientific research on cat intelligence: the past decade has seen an explosion in studies of the mental abilities of dogs, but cats, popular subjects for such studies in the 1960s and 1970s, have subsequently been eclipsed by "man's best friend"—simply because dogs are easier to train.

Some recent studies have focused on cats' understanding of the way the world around them works—their grasp of physics and engineering, if you will. Cats may seem entirely *au fait* with their surroundings, but their capacity to translate the phenomena they encounter into mental pictures evolved when they were wild animals; it has not caught up with humankind's manipulations. I realized this firsthand when I noticed that my cat Splodge always inspected the fenders of cars parked outside my house. Sometimes after sniffing he looked around nervously, and I guessed that he must have picked up a scent mark from another cat, probably deposited a few hours previously when the car had been parked elsewhere, miles away. This happened day after day, month after month, but Splodge, who was otherwise a perfectly intelligent cat, never seemed to understand the possibility that the scent marks might have arrived already on the car: he always seemed to presume that they belonged to an unfamiliar cat that must have just arrived in our neighborhood. In nature, scent marks stay where they've been left, so there would be no need to evolve an understanding that scent marks might move with objects on which they've been deposited.

We have little research on cats' comprehension of physics, but one recent experiment has confirmed that it may be extremely rudimentary. Scientists trained pet cats to retrieve a food treat from beneath a mesh cover by pulling on a handle connected to the treat by a string (see illustration on page 142). Many of them learned to do this quite easily, giving the impression that they "understood," as you or I would without thinking twice, that the handle was connected to the food by the string. However, we can also explain this behavior by simple operant conditioning—pull the handle and a food reward arrives—in which the handle-pulling is, so far as the cat is concerned, just an arbitrary action. The researchers exposed the cats' lack of understanding of the connection by adding another string and handle alongside the first one; the crucial difference, which the cats could easily have seen, was that only the first handle was connected to the food. Although the cats could easily have seen that this was the case, they continued to pull on the handles, but were unable to predict which handle would produce the reward, showing that for them, the handle was an arbitrary object, not physically linked to the piece of food. Unsurprisingly, the cats performed equally badly when the strings were crossed over.[15] Extrapolating from this experiment, it seems likely that cats, unlike crows or apes, are mentally incapable of learning to use tools.

Their inability to understand that a piece of string can physically link two other things together highlights just how different cats' minds are to our own. We not only find such an idea obvious, we routinely extrapolate it to other situations in which the physical connection is much less obvious, such as the electronic connection between the cursor on the screen in front of me and the (computer) mouse in my hand. Cats fail at the first hurdle, the comprehension of

Cats appear not to understand that one string has food on the end, and the other doesn't

a physical connection between three objects, and their inability to perform such manipulations is not, as some would jokingly have it, simply due to their lack of opposable thumbs.

Yet cats do have a sophisticated understanding of three-dimensional space, as we would expect of an opportunistic hunter. They probably have little need for this when indoors, instead relying on what are known as egocentric cues—"This is where I turn left," "This is where I turn right," "This is where I jump up." However, when outdoors and in familiar territory, cats can take shortcuts, showing that during their earlier explorations they have constructed a mental map—"The last time I caught a rat, I first went to the oak tree and then turned left down the hedge, so this time I will go diagonally across the field and through the hedge; I already know that the rats' hole is just beyond the hedge." They are capable of using this information efficiently; given a choice of routes to a destination that cannot be seen from where they are, cats will pick the shortest one. Likewise, as many people also do, they prefer a route that starts out in roughly the right direction; a slightly shorter route that involves initially walking in the wrong direction is usually shunned.

As hunters, cats should be able to work out where objects that have gone out of sight are likely to be. No wild cat would give up hunting a mouse immediately after it disappeared from sight under the mistaken impression that it had ceased to exist. As expected, cats do seem to remember where prey has disappeared, although they store this information for only a few seconds, in "working" memory; not until the cat actually makes contact with the prey does the memory become longer-lasting. Presumably, it is not worthwhile for a cat to continue to search a particular location for highly mobile prey for much longer than this; by that point, the prey has either made its escape or gone to ground.

Scientists recently demonstrated that cats do indeed remember the last place they saw a mouse, rather than merely keeping their eyes fixed on it or even just heading in that general direction. In the apparatus illustrated, scientists allowed the cat to watch a food treat being pulled behind a small barrier on a piece of cord, through the

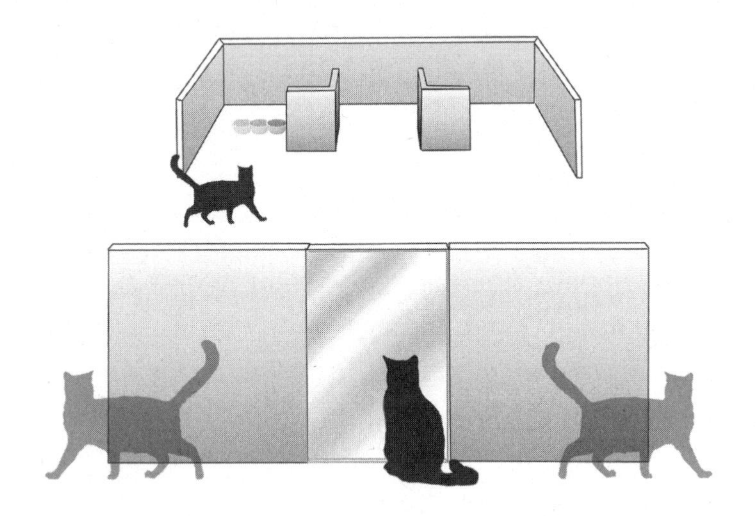

Cats sometimes take the most direct route to the place where food has disappeared (left), but other times they seem to deliberately take a more indirect route (right), as if they were hunting and wanted to confuse their prey.

transparent section of the screen. The cat was then allowed to walk into the apparatus, but—as the remainder of the screen was opaque—this action temporarily blocked the cat's view of the location of the food. Nevertheless, the cat usually picked the right place to look for the food. Interestingly, many of the cats tested sometimes went the long way around—they set off and entered the apparatus from the "wrong" side, but then immediately crossed to the correct side to retrieve the food. This is similar to a common hunting tactic: if cats are in hot pursuit of a mouse or rat, they will often briefly take a more roundabout route, possibly to make their prey think that they have mistaken where it has hidden.[16]

The cat's mental abilities are specifically tuned to their hunting lifestyle, rather than being part of a more general spatial intelligence. Cats perform remarkably badly in tests designed to track the development of such abilities in human infants. Many children of eighteen months can understand that if an object is hidden in a container, and then the container itself is hidden, the object ought to be in the container when it reappears. If they then find that it isn't, they look next in the place where the container disappeared. They not only understand that an object they have seen must still be somewhere even when they can't see it, but can also use their imagination to guess where it might be. Cats cannot do this at all, probably because it is not a situation that their ancestors encountered while hunting. Mice hide, certainly, but they do not hide inside objects that are themselves capable of moving around.

The cat's ability to reason seems limited, especially when determining cause and effect. They rely on simple associations built up through conditioning, and can easily be "fooled" by our manipulations of their surroundings, which must seem arbitrary in their view of the world (as if they were thinking "How did that bag arrive in the middle of the kitchen floor?" or "Why does my owner talk into that thing in her hand?"). Nevertheless, it is entirely possible that scientists have yet to design experiments that allow cats to demonstrate their true abilities. It could be that the small number of situations in which cats have been tested happen to be those in which their evolution has favored reliance on simple learning and short-term, rather than long-term, memory. Scientists studying canine intelligence, which has received considerable attention over the past two decades, have only recently begun to find ways of testing dogs that fit their particular way of interpreting the world. Cats, with their enigmatic reputation, may still be hiding the true extent of their brainpower.

Cats are masters at concealing their thoughts, and are even better at hiding their emotions. Several cartoons show an array of cats with identical expressions, each labeled with a different emotion, ranging from "Frisky" to "Content" to "Sad," and in one instance, with irony, "Alive." One version that I treasure, by British cartoonist Steven Appleby, has no less than thirty cat faces, of which twenty-nine have identical expressions (the captions range from "About to do nothing at all" to "Slightly irritated but concealing it well"); the thirtieth, "Asleep" differs from the others only by its closed eyes.[17]

Biology provides good reasons why most animals keep their emotions to themselves. Dog owners may find this idea absurd, since both dogs and humans express their emotions spontaneously. Indeed, we must often suppress our feelings when social mores dictate. Nevertheless, we humans have evolved a very sophisticated ability to detect tiny flickers of emotion in others, telltale signs that help us predict what that person is likely to do next. Dogs, in their way, are similar to us in this respect; not only have they evolved the ability to guess our intentions from our body language, but they also express their emotions openly, partly because we generally respond in ways that are beneficial to them—backing away when they growl, or giving them a pat if they're wagging their tails. Dogs and humans are both social species that usually live in stable groups, and such stability means that emotional honesty is unlikely to be penalized.

Cats are descended from a species with a solitary lifestyle, and therefore much of their behavior is guided by the need to compete, not to collaborate. In the wild, a male cat will live alone. The only way he can be sure of leaving any descendants is first to convince a female to accept him as a mate, and second to convince any rival males to back off. Macho behavior, laced with a generous portion of bluff, is therefore essential to their success. Although female domestic cats do cooperate when raising kittens, this habit, which may have evolved during domestication, seems to have had little effect on their capacity to express their emotions.

Historically, scientists have changed their mind several times over whether animals' emotions should be used when discussing their behavior. In the nineteenth century, researchers often ascribed human emotions to cats. For example, in his 1886 book *Animal Intelligence*, physiologist George S. Romanes wrote:

> The only other feature in the emotional life of cats which calls for special notice is that which leads to their universal and proverbial treatment of helpless prey. The feelings that prompt a cat to torture a

captured mouse can only, I think, be assigned to the category to which by common consent they are ascribed—delight in torturing for torture's sake.[18]

By the beginning of the twentieth century, such anthropomorphism had been abandoned, and the guiding principle in animal psychology was Morgan's canon: "In no case may we interpret an action as the outcome of the exercise of a higher psychical faculty, if it can be interpreted as the outcome of the exercise of one which stands lower [that is, simpler] in the psychological scale."[19] For a while, indeed, scientists conceived of animals as if they were robotic stimulus-response machines, leaving no room for any consideration of emotion. Recently, however, we have come to realize that it is difficult to explain some animal behavior without invoking the idea of emotion. In addition, MRI scanning has enabled us to see where in the human brain emotions are generated, and the simpler, "gut-feeling" emotions occur in parts of our brains that we share with other mammals, including cats.

The current view holds that emotions are a necessary component of the mechanisms that drive animal behavior—and indeed our behavior as well. They can be prompts to shortcuts, enabling our brains to choose the best response to a situation when fast action is called for. In this, cats are no different from ourselves. A cat that sees another, larger, unfamiliar cat approaching will immediately become alert, crouch down, and prepare to flee. The anxiety it feels at seeing a potential adversary enables it to take these actions immediately, without having to think the situation through and evaluate all possible strategies and outcomes.

Emotions also explain spontaneous and apparently functionless behavior. Kittens engage in play during most of their waking moments, and explaining why they do so is not straightforward. In the wild, play is a mildly risky activity, exposing the kittens to danger and possibly attracting the attention of a predator—surely it seems safer for kittens to stay quietly in their nest and wait for their mother to return with food. Furthermore, kittens do sometimes nip each other when playing, but this doesn't seem to put them off playing with that same kitten again, which it should do if they are simple stimulus/response machines.

The simplest explanation for why kittens feel they ought to play, and why they continue to play even following a slight mishap, is that play is *fun*. Neuroscientists have found in young rats that when they play, the neurohormone profile of the brain changes. Moreover, these changes are not the consequences of play, they appear to be its cause: they occur as soon as the rats are given the signal that it's time to play. As such, the mere sight of a sibling ready to play is likely enough to make a kitten want to join in, because the brain signals "fun" before the playing has even begun.

Of course, hormones are not the same as emotions, but changes in certain hormones are often a sign that emotions are being experienced. We are all aware of the racing heartbeat, hyperventilation, hyperalertness, and sweaty palms induced by adrenalin or epinephrine, the "fight-or-flight" hormone associated with feelings of fear and panic. Some of us are familiar with the elation we sometimes feel after strenuous exercise, caused by the release of endorphins and other hormones into the brain. Although not all hormones are so closely connected to emotions, many are, and can provide an indicator of immediate emotion or underlying mood.

We can therefore conceive of animal emotions as manifestations of the brain and the nervous system and their associated hormones, sometimes enabling decisions to be made quickly, other times directing learning. Sometimes, information coming into the brain from the cat's senses has to trigger an immediate reaction. A cat that slips while walking along the top of a fence must correct its balance immediately: the emotional panic that almost certainly follows will help train the cat to be more careful next time. On other occasions, the emotion triggers the behavior. Indeed, the sight of its owner coming home will cause a cat to feel affectionate, and as a consequence it will raise its tail upright and begin to walk toward her.

Some people who don't like cats—and even some who do—would claim that love, especially towards its human owner—is not part of the cat's behavioral repertoire: as the popular adage goes, "Dogs have masters, cats have staff." True, the average cat does not outwardly demonstrate love for its owner in the same way as, say, a Labrador retriever would. Still, that tells us little about what is going on inside the cat's head.

In the animal world, lavish outward displays of emotion are usually manipulative. Consider, for example, the incessant chirping of baby birds in a nest, each essentially calling out, "Feed me first, feed me first!" Evolution has ensured that where such displays occur, they work: the baby bird that makes the least fuss is ignored by its parents, which often produce more offspring than they can comfortably feed, and it may perish as a result. Wildcats are solitary and self-sufficient except for a few brief weeks at the beginning of their lives, and so have little need for sophisticated signals. Although domestic cats now depend on us for food, shelter, and protection, they have not done so for long enough to have evolved the average dog's effusive greeting. That doesn't mean that cats are incapable of love, merely that their ways of showing love are somewhat limited.

Cats become extravagantly demonstrative only when they are angry or afraid. A fearful cat either will make itself look as small as possible by crouching, and then slink away, or, if it judges that running away may provoke a chase, will make itself look as large as possible, arching and raising the fur on its back. The raw emotion does *not* therefore provoke an automatic and invariable reaction; instead, the brain selects the more appropriate response based on other information available to it.

The angry cat will not only try to look as large as possible, but will also stand head-on toward the threat (usually another cat) with its ears forward, either yowling or growling loudly and lashing its tail side to side. We may be inclined to read these postures as expressions of emotion, but they are fundamentally expressions of intention, as well as attempts to manipulate the animal the cat is confronting.

Even when a cat is demonstrably trying to manipulate his opponent's behavior, such manipulation need not be conscious. We can explain the cat's behavior by its adhering to a set of rules that have served its ancestors well; which have, in the past, achieved the best possible outcome from an aggressive encounter. We must not forget, however, that the bluff is being directed at an animal with similar brainpower, and therefore evolution has also favored animals capable of "mindreading" the other's intentions. One cat confronting another will expect one of two possible reactions: either a sign that the other cat is fearful and will probably back down, or that it is not fearful and so a fight is imminent. In this way, the behavior, whether indicating fear or anger, becomes ritualized: any cat that adopts neither posture, or does something different, is likely to be attacked.

To explain their behavior, we must thus allow that cats feel joy, love, anger, and fear. What other emotions do they possess? Can they feel the full gamut of emotions that we can? To answer these questions, we must consider which are likely the product of human consciousness, and therefore unknown to animals.

People do not agree on the range of emotions their cats can feel. A 2008 survey of British cat owners revealed that almost all think their cat could feel affection, joy, and fear.[20] Nearly one-fifth of these owners—perhaps those who own very timid cats—were unsure whether anger is within the cat's repertoire.

We all know the old saw "Curiosity killed the cat," and indeed, most owners acknowledge the cat's characteristic nosiness. The original version of this proverb, from its first appearance in the sixteenth century until the end of the nineteenth, was "Care killed the cat"—"care" in the sense of worry, anxiety, or sorrow. Apparently, the idea that cats could become so anxious that they could even die from it was once common (and is now being revisited by veterinarians). Despite this, about a quarter of the owners surveyed in 2008 thought their cat incapable of feeling anxiety or sadness.

If asked, scientists would now agree that the old version of the proverb contained a germ of truth: anxiety does constitute a serious and real affliction for many cats. Anxiety, if simply defined as a fear of something that is not currently happening, has a reliable basis in physiology. Some anti-anxiety drugs developed for humans have also been found to reduce symptoms of anxiety in cats, so although we

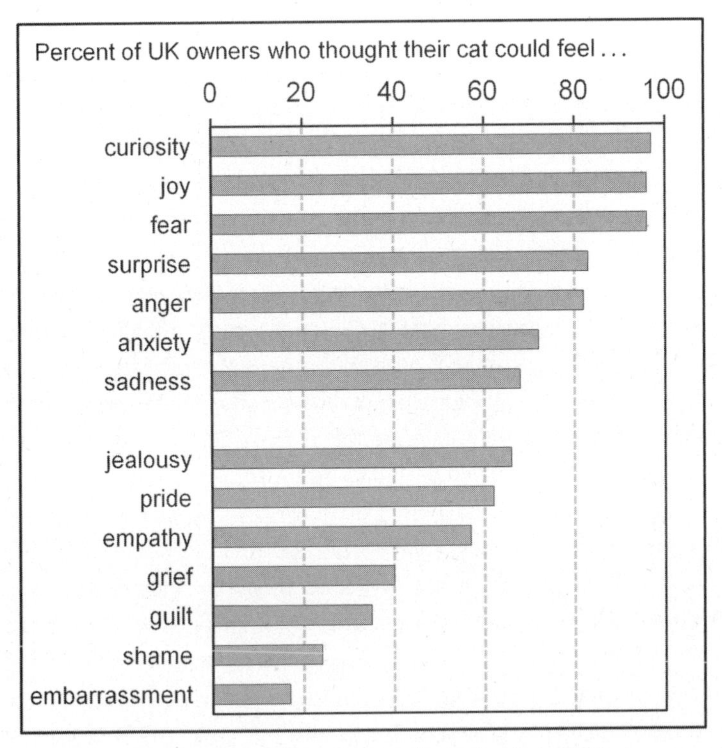

Owners' opinions about their cats' emotional capabilities

cannot be sure that cats experience anxiety in precisely the same way we do, we know that they feel something similar.

The most common cause of anxiety in cats is probably the worry that their territory is likely to be invaded by other cats in the neighborhood, or even by another cat in the same household. When I surveyed ninety cat owners in suburban Hampshire and rural Devon in 2000, they reported that almost half of their cats regularly fought with other cats, and two out of five were fearful of cats in general. My colleague Rachel Casey, a veterinary surgeon specializing in cat behavioral disorders, regularly diagnoses anxiety and fear as the main factors driving cats to urinate and defecate indoors, outside the litter tray. Some cats spray the walls or furniture with urine, possibly to deter other cats from entering their owner's house believing it to be cat-free; others find the point in the house farthest from the cat-flap and urinate there, seemingly terrified of attracting the attention of any other cat. Some will defecate on the bedsheets, desperately trying to mingle their own odor with their owner's to establish "ownership" of the core of the house. When the conflict is between two cats that live in the same house, one may spend much of its time hiding, or obsessively groom itself until its coat becomes patchy.[21]

The stress of being forced to live with cats it does not trust can often be severe enough to affect the cat's health. One illness now known to be closely linked to psychological stress is cystitis, referred to by veterinarians as idiopathic cystitis because no disease or other medical cause is apparent. As many as two-thirds of cats taken to vets for urination problems—blood in the urine, difficult or painful urination, urinating in inappropriate places—have no obvious medical problems, other than inflammation of the bladder and intermittent blockage of the urethra by mucus thereby displaced from the bladder wall. The factors triggering such episodes of cystitis are therefore psychological, and research has identified conflict with other cats living in the same household as being perhaps the most important of these. Less easy to quantify, but possibly just as important, are conflicts with neighbors' cats: certainly cats prone to cystitis usually run away from cats they meet in their own gardens, rather than standing up to them, which suggests that they find contact with other cats particularly stressful. Idiopathic cystitis is less common in female cats than in males: the conventional medical explanation is that the tube leading out of the bladder, the urethra, is generally narrower in males and therefore more prone to blockage. However, tom cats are generally more territorial and less sociable than females, so the latter may find it easier to resolve or avoid conflicts with other cats before stress affects their health.

Colleagues of mine at Bristol University Veterinarian School documented one case involving a five-year-old male cat that was having great difficulty urinating, and when he did, his urine was bloody. He also groomed his abdomen excessively, but otherwise was entirely healthy. The cat was one of six within the household, but was friendly with none of the other cats. Moreover, cats from neighboring households had recently attacked him. His symptoms gradually disappeared once his owners had implemented the changes recommended by the clinic: his own exclusive area within the house, his own food bowl, and his own litter tray that the other cats could not access. At the same time, his owners blocked his view of the garden by covering over the glass at the bottom of the windows in his part of the house, so he could not see any cats that were coming into the garden. Six months later his symptoms returned, but on investigation, it turned out he had been accidentally shut in with the other cats a couple days previously. His owner vowed never to allow this to happen again, and the cat soon recovered.[22] Anxiety, a useful emotion if experienced for a few minutes, can become the bane of a cat's existence if prolonged for weeks or months, leading to chronically elevated levels of stress hormones, (presumably) a nagging and ever-present sense of dread, and eventually to deterioration in health.

In the same survey, cat owners were also asked about their cats' more complex emotions—jealousy, pride, shame, guilt, and grief. Almost two-thirds believed that their cat could feel jealous or proud. Only shame, guilt, and grief were ruled out by a majority of the owners.

Basic emotions such as anger, affection, joy, fear, and anxiety are "gut feelings" that appear spontaneously. The most primitive part of the cat's brain produces these emotions—the same part that evolved hundreds of millions of years ago before there were any mammals, let alone cats. More complex emotions, such as jealousy, empathy, and grief, require the cat to have some understanding of the mental processes of animals other than themselves, and hence psychologists sometimes refer to them as *relational* emotions.

Take jealousy, for example. When we experience jealousy, we are not only aware that whoever we are feeling jealous of is another human being, we can also guess what that other person is feeling; we have what psychologists call theory of mind, the idea that other humans have their own thoughts that may be different from ours. We are also capable of becoming obsessively jealous, thinking about the incident that triggered the original feeling long afterward, even when the person we were jealous of is no longer present. We have little evidence that cats have either the brainpower or the imagination to do either of these.

Cats undoubtedly recognize other cats as cats, and can evidently react to what they see them *doing*. However, even dogs, which are much more highly evolved socially than cats, show no evidence of understanding what other dogs are *thinking*, so it's unlikely that cats can either. Moreover, cats seem to live in the present, neither reflecting on the past nor planning for the future. Still, at its heart, jealousy is an emotion first experienced in the here and now; it does not require the cat to understand what its rival is thinking, or even that it is capable of thinking at all. All that jealousy requires is that the cat merely perceive that another cat is getting more of something than it should. Thus, cats are almost certainly capable of feeling jealous, even if not quite as demonstratively, or as commonly, as dogs. Although not something that my cats have ever indulged in, countless owners have regaled me with stories of how one of their cats would always intervene when they tried to stroke the other.

Many people think that cats are capable of grief, because they behave oddly when another cat that they have known disappears. What they actually feel is probably a temporary anxiety, which disappears once all traces of the missing cat have disappeared. A mother cat may search for her kitten for a day or two after it has been homed. She probably has a memory of that kitten, and may even count the remaining kittens to check that one is missing. This behavior would be the same if that kitten had temporarily gotten lost; in the wild, it would be in the mother cat's interest to seek it out and continue to look after it until it was old enough to become independent of her. She cannot "know" that it has gone to a good home where it will be well cared for, as nothing in her evolution will have prepared her to embrace that concept. For a few days, the mother is reminded of that particular kitten by the lingering traces of its individual odor, the kind of cue that is often meaningless to us. We know that the kitten has gone because we can no longer see or hear it.

Once the kitten's odor has faded below her threshold, the mother cat probably forgets all about the departed kitten. While she can still smell it, she may well feel an anxiety that drives her to continue to look for it. This, however, is not the same as grief.

Emotions such as guilt and pride would require cats to possess a further level of cognitive sophistication, the ability to compare their actions with a set of rules or standards that they have worked out for themselves. When we feel guilty, we compare the memory of something we've just done with our sense of what is wrong. Such feelings are sometimes referred to as self-conscious emotions, because they require a degree of self-awareness to be experienced. So far, science has yet to reveal any evidence for self-awareness in cats, or even in dogs. Dogs are widely believed to display a "guilty look" when their owners discover that they've done something that's forbidden, but a clever experiment has shown that this is all in the owners' imagination.[23] The researcher asked the owners to command their dogs not to touch a tempting food treat, and then leave the room. Then, unbeknown to the owners, she encouraged some of the dogs, but not others, to eat the treat. When the owners came back into the room, they were all told that their dog had stolen the treat—whereon all of the dogs immediately began to look guilty, whether or not they actually had something to feel guilty about. The "guilty look" was nothing more than each dog's reaction to their owners' body language, which had changed subtly as soon as they had been told of the dog's misdeed, whether real or invented. If dogs' "guilty looks" are a figment of their owners' imaginations, it follows that they—and by extension, cats—are incapable of feeling guilt. The same is probably true for pride, but no scientist seems to have studied it in any relevant animal.

Cats' emotional lives are more elaborate than their detractors would have us think, but not quite as sophisticated as the most ardent cat-lover would probably like to believe. Unlike dogs, cats hide their emotions—not primarily from us but from one another, a legacy of their evolutionary history as solitary, competitive animals. We have every reason to believe that they possess the basic range of emotions, the gut feelings, shared by all mammals, because having these empowers them to make quick decisions, whether that is to run away (fear), play with a ball of string (joy) or curl up on their owner's lap (love). However, cats are not as socially sophisticated as dogs are: they are undoubtedly intelligent, but much of that intelligence relates to obtaining food and defending territory. Emotions that relate to relationships, such as jealousy, grief and guilt, are probably beyond their reach, as is the ability to comprehend social relationships with any great sophistication. This leaves them ill-equipped for the demands of living closely with other cats, as domestication has progressively required them to do.

CHAPTER 7

Cats Together

Cats can be very affectionate, but they are choosy about the objects of their affection. This apparent fastidiousness stems from the cat's evolutionary past: wildcats, especially males, live much of their lives with no adult company, and regard most other members of their species as rivals rather than as potential colleagues. Domestication has inhibited not only the wildcat's intrinsic distrust of people, but has also tempered some of their wariness of other cats.

The bond between cat and owner must have its origins in the bond between cat and cat; such behavior has no other plausible evolutionary source. Although the cat's immediate ancestor, the wildcat, is not a social animal, adult felids of other species, such as lions, do cooperate. As such, cats of any species could possibly become more sociable in the right conditions. We may find the source of the domestic cat's affection for its owners through a brief survey of the social life of the entire cat family.

Both male and female tigers are solitary, exemplifying the pattern that almost all members of the cat family, big and small alike, live alone. The females hold non-overlapping territories, which they defend from one another; each territory is large enough to provide food not just for that female, but also for the litters of cubs she rears. Young males are usually nomadic, and when they become mature, they try to set up their own large territories. These will contain far more prey than the male will ever need to satisfy his hunger, but that is not its purpose. The male is trying to achieve exclusive access to as many females as possible: especially successful males may hold territories that overlap those of up to seven females.

Cheetah males, especially brothers, are a bit more sociable than tiger males. Female cheetahs are as solitary as female tigers, but male cheetahs sometimes band together in twos or threes to seduce females, many of which are migratory, as they pass nearby. Even though only one of the brothers will be the father of the resulting litter, biologists have shown that over his lifetime each brother will father more cubs than if he had tried to attract females on his own. Male cheetahs do occasionally try to hunt together as well, but are rarely successful, apparently lacking the skills to coordinate their efforts.

The best-known exception to the standard felid pattern is the lion, the only member of the cat family to have several males and several females living together. In Africa, the lion pride is usually constructed from one family of female lions, with the males originating from a different family (thereby preventing inbreeding). While still young, related males band together, sometimes adding unrelated males to swell their ranks, until their numbers are sufficient to challenge and eject the resident males from a pride. Once they have taken over the pride, they may kill all the cubs, and by doing so bring all the females into season within a few months. The males must then keep control of the pride until the females have given birth to their progeny and raised them to independence. For their part, the females not only have to raise the cubs, but they also do most of the hunting, while the males do little but protect the females from other bands of males. Thus, the superficial image of lion society as generally harmonious is something of a myth; rather, it is the result of tension between the benefits of cooperation and competition, with each individual lion employing tactics that maximize its own breeding success.

Biologists still debate why lions live in such groups. In India, lions are often solitary animals, males and females meeting only for mating, so members of this species can evidently choose whether or not to live together. Although the females in a pride generally hunt together, they do not cooperate as selflessly as their "brave" image suggests: if the prey is large and potentially dangerous, the more experienced females usually hold back and let the younger, more impetuous lionesses take the risks. The main benefit of numbers, and especially the presence of the fierce males, may come after the kill, when the valuable meat must be defended from other animals, particularly hyenas.

Scientists once considered lions and cheetahs the only social cats, and added the domestic cat to that exclusive list only recently. It had long been clear that wherever there was a regular source of suitable food, a group of feral cats would spring up, but these groups were originally considered mere aggregations of individuals that had somehow agreed to tolerate one another, as happens when animals of many species come together to drink at a waterhole. Cat breeders also knew that their queens would sometimes suckle one another's litters, but scientists dismissed this behavior as the result of the artificial conditions under which humans usually keep pedigree cats. In the late 1970s, however, David Macdonald's documentary about cats on a farm in Devon, England, showed that this was in fact natural behavior—that free-living females, especially related females, will spontaneously cooperate to raise their kittens together.[1]

At the start of the study, the colony consisted of just four cats: a female, Smudge; her daughters, Pickle and Domino; and their father, Tom. When not in the farmyard, they kept to themselves—domestic cats, unlike lions, do not hunt together—but when their visits coincided, they usually curled up together in apparent contentment. They evidently regarded the farmyard and the food and shelter it provided as "theirs," since the three females would join together to drive away three other cats that lived nearby—a female, Whitetip, her son Shadow, and her daughter Tab. Tom, though, was aggressive only toward Shadow, presumably regarding him as a potential rival, whereas he might need to court the two females should he lose his own "pride" at some point in the future.

Pickle and Domino were the first to reveal how female cats come together to help one another. Early in May, Pickle produced three kittens in a nest in a straw stack. For the first couple weeks, she looked after her kittens on her own, just as any other mother cat would. Then, suddenly, her sister Domino appeared in the nest and gave birth to five more kittens—ably assisted by Pickle, who helped with their delivery and with cleaning them up. Then, despite their difference in ages, all eight kittens settled down together and were nursed and cared for indiscriminately by both their mothers. Sadly, all eight kittens subsequently succumbed to cat flu, a common scourge of litters born outdoors in the UK. However, when their grandmother Smudge produced a single male kitten a few weeks later, the aptly named Lucky, both Domino and Pickle helped with caring for him, including playing with him and bringing back mice they had caught, to save Smudge leaving him alone while she went out hunting for herself.

Subsequent studies have revealed that such cooperation between related female cats is the rule, not the exception. I observed this in

Domino and Lucky playing

my own home when my cat Libby gave birth to her first litter: Libby's own mother, Lucy, shared their care, grooming them and curling around them to keep them warm. Indeed, when they grew old enough to move around the house, Libby's kittens generally preferred their grandmother's company to their mother's.

Cat society is based on females from the same family. In feral or farm cats, it rarely involves more than two (sisters and their kittens) or three generations (mother, daughter[s], and kittens). However, we see little indication that the participants are consciously helping one another. Rather, many female cats, especially those with kittens, simply seem not to distinguish between their own offspring and those of other cats with whom they are already friendly; in the wild, these are most likely their own daughters or sisters—cats they have known and trusted all their lives. Some mother cats that have recently given birth will accept almost any kitten they are presented with, and some humane organizations make use of such cats as wet nurses for motherless litters.

Larger colonies of feral cats usually consist of more than one family, and while these families continue to cooperate among themselves, they also compete with one another. The size of a cat colony is determined by the amount of food available on a regular basis, and where this is plentiful—for example, in a traditional fishing village where the catch is processed *in situ*—colonies can build up until several hundred cats are all living in close proximity to one another. Cats are prolific breeders, and numbers can grow quickly until food becomes scarce, at which point the cats on the edge of the colony will either leave or succumb to disease brought about by malnutrition.

Each family group strives to monopolize the best sites for dens in which kittens can be born, and to stay as close as possible to the best places to find food. However, as the most successful families grow, tensions increase among its members, even when there is enough food. Cats appear to be incapable of sustaining a large number of friendly relationships, even when all their neighbors are close relatives. Squabbles break out, and eventually some members of the family are forced out—and because all the prime space in the colony will already be occupied by other cats, they may have to find a place alongside the new arrivals on the very fringes of the colony, where pickings are slim.

Cat colonies are therefore far from being well-regulated societies: rather, they are spontaneous gatherings of cats that occur around a localized concentration of food. If the food supply is limited, a single family may monopolize it. When food is especially plentiful, several families compete for the best and biggest share, although warfare between the various clans is generally conducted through threats and careful skirting around one another, punctuated only occasionally by overt violence. In such situations, being able to call on family to assist is essential to holding on to prime territory: a female cat on her own, especially if she has kittens to feed, is unlikely to thrive.

The cooperation within families that undoubtedly occurs in larger colonies is based on the same bonds of kinship that form in much smaller colonies consisting of a single family. Cats seem incapable of forming alliances between family groups, unlike, for example, some primates; negotiation skills of this sophistication lie beyond their capabilities.

Biologists are uncertain about the precise origins of these family ties. They could be accidental, caused by female cats' inability to distinguish their own kittens from others. Looking back to their wildcat ancestors, every female holds her own territory, which she defends against all other females, so the chance that two litters could ever be born in the same place is virtually zero. A female cat, wild or domestic, probably follows a simple rule of looking after all the kittens that she finds in the nest she has made: she sees no need to sniff each kitten carefully to check that it is not an interloper before settling down to nurse it. However this is unlikely to form the sole basis for cooperation among adult cats. The thousands of generations over which domestic cats have evolved away from their wild ancestors provides enough time for more sophisticated social mechanisms to have evolved.

Cats' social behavior probably started to evolve as soon as humankind's invention of food storage first made concentrated sources of food available. Any cats that maintained their natural antagonism toward all members of their own species could not have exploited this new resource as efficiently as those able to recognize their relations, and both give and receive help from them.

Biologists distinguish two different ways in which cooperative behavior can be beneficial to both parties. One is reciprocal altruism, continuing to do favors only for those who have done favors for you. This could theoretically take place between any two animals that live near each other, irrespective of whether they were related. However, if they are related, a second reason why it would be a good idea to cooperate is kin selection. Cats that are sisters share half their genes, a much greater proportion than two unrelated female cats would. Their kittens, even if fathered by different tomcats, each share a quarter of their genes with their aunt. Thus, whichever of those kittens goes on to have kittens of his or her own (it's an inevitable fact of life that not all of them will), each kitten shares genes with both their mother and their aunt.[2] Since neither sister knows which of their offspring is most likely to survive to maturity, they should, all other things being equal, try to raise both litters. Thus, the genes that favor cooperation between cats that are related can flourish, at the expense of rival genes that promote antagonism, even between sisters.[3]

Reciprocal altruism and kin selection are useful mechanisms preventing selfish behavior, but cooperative behavior itself will evolve only if its benefits outweigh its costs. For the first domestic cats, the initial advantage of living in family groups would have been that the abundant food—whether prey infesting food stores, scraps provided by people, or a mix of the two—could be shared without constant strife. However, putting several litters into one nest means that if one kitten gets sick, they all do; this can be fatal, as Domino and Pickle's experience illustrates. When in 1978 South African scientists introduced a virus to exterminate cats that were causing havoc among the ground-nesting seabirds on Marion Island in the Indian Ocean, cats that had retained their wild ancestors' habit of nesting on their own mostly survived, but family groups perished. Elsewhere, this disadvantage may be outweighed by the benefit of having several mothers on hand to protect kittens from predators: a solitary mother must leave her litter from time to time to find food, or her milk will dry up.

Two or more mother cats that pool their kittens can guard them much more effectively than can a lone mother, who has to leave her kittens alone in order to hunt. This benefit must outweigh the increased risk of disease wiping out the whole litter, or the ability to cooperate like this would never have evolved. Domestic cats living among humans have probably always had two main enemies: stray dogs and other cats. Twenty years ago, while on holiday in a Turkish village with my wife and our youngest son, our apartment was visited by a heavily pregnant calico stray we christened Arikan. After disappearing for a couple of days, she returned, much thinner and starving hungry, so we knew that she must have given birth nearby. We found a supermarket that sold cat food (receiving puzzled looks from the cashier, who seemed not to expect tourists to make such a purchase). When we followed Arikan after she ate, she disappeared into some derelict farm buildings farther up the road. After that, we fed her morning and evening, until one night we were awoken by pitiful wailing: Arikan was outside our door, carrying a dead kitten that showed signs of having been mauled. Arikan immediately ran off, and moments later returned with another dead kitten, which she deposited alongside the first on our doorstep.

A dog may conceivably have been the culprit. Small packs of "latchkey" dogs did roam the village in the evenings, badgering diners at outdoor restaurants for table scraps and chasing cats. If a pack of dogs found a cat's nest, they would probably dismantle it, but possibly not eat the kittens—although on one occasion, I did witness a dog playing with a dead kitten. Under other circumstances, dogs have been effective predators of cats: in Australia, dingoes—originally domestic dogs, gone wild—keep feral cats in check, enabling local small marsupials to thrive.[4] Still, despite a long history and reputation for being the cat's prime nemesis, dogs were likely not to blame for these kittens' deaths.

Because Arikan's loss occurred at night, the culprit was more likely another cat, the dogs having returned to their owners' homes at dusk. Infanticide is a regular occurrence in lions, but only a few instances have been documented in domestic cats.[5] Male lions kill cubs that are not their own because doing so brings the mother lion back into season almost immediately; otherwise, the males must endure a nineteen-month interval between births, and by that time those males might have lost control of the females.

Female cats are often ready to mate as soon as their kittens are weaned, or earlier if they do not survive infancy, so the intruding male cats that have carried out all recorded instances of infanticide probably gained little from their cruel act, at least in terms of increasing their opportunities for mating. Infanticide appears to be most common in small, single-family cat colonies on farms, rather than—perhaps surprisingly—in the large multi-group colonies where aggression is much more generally evident. In those larger colonies, females often mate with more than one male, thereby making it much more difficult for those males to work out which kittens are theirs and which are not. In this way, males should kill only the kittens of females they can be sure they have never met before.

If the mother is present, she defends her kittens with all her might, so males are wise to target unguarded litters. A pooled litter with two or more females in attendance will be better protected against marauding males than litters kept in separate nests, even though they are less well-isolated from infections. Perhaps the unfortunate Arikan had no surviving sisters with whom she could have joined forces.

Family life provides opportunities for cats to learn from one another, rather than working everything out for themselves. As we have seen, mother cats instruct their kittens on how to handle prey by bringing it back to them. We have no evidence that she actively *teaches* them: she simply provides them with opportunities to learn what prey is like, but in the safe environment of the nest. Also, although kittens naturally pay great attention to everything their mothers do, it is unlikely that they deliberately imitate her. True imitation involves complex mental processes; the animal must first know what the relevant actions are, and then translate what it has seen into movements of its own muscles. Because we ourselves find it easy to imitate, we tend to assume that other animals can do the same, but research indicates that true imitation—deliberate copying of another animal's actions—may be restricted to primates.

We know of simpler ways that kittens learn from their mothers, without directly imitating her actions. Instead, the mother draws the kittens' attention to an appropriate target, and they then direct their own instinctive behavior toward it. In a 1967 experiment, scientists demonstrated that mothers teach their kittens best when challenging them to perform an arbitrary task to get food. The experiment consisted of allowing a kitten to explore a box that had a lever sticking out of one wall.[6] The kitten would usually ignore the lever, apart from giving it a sniff, unless it had watched its mother pawing the lever and being rewarded with food, in which case it paid great attention to the lever and quickly learned that pawing it produced a reward. Usually, kittens will spontaneously pat an object that moves, but only rarely pat something that appears inanimate and fixed. As other experiments have shown, that both mother and kitten ended up pawing the lever was probably not due to direct imitation—the lever was designed to work best when pawed—but instead to the kitten's realization that the lever it had seen its mother manipulating looked the same as the one in its own box, and was therefore something it should investigate closely. Kittens invariably use their paws to investigate something novel, so they didn't need to imitate their mother's actions; they simply did what came naturally. In a similar experiment, other kittens were allowed to watch unfamiliar female cats performing the task. These kittens either took a long time to learn to paw the lever or didn't learn it at all, showing that it is specifically the mother that kittens feel comfortable watching. They are probably just too inhibited by the presence of an unfamiliar adult cat to learn anything.

Mother cats can pass some of their hard-won expertise on to their kittens by providing them with novel experiences, but this is equally easy for solitary mothers and those that live socially, in families. Can adult cats benefit from living socially by learning from one another? We know little about this intriguing possibility, but one obscure experiment done more than seventy years ago suggests that they can.[7] Several six-month-old cats were given the opportunity to obtain a bowl of food placed on a turntable out of their reach, as shown in the illustration. A ratchet under the turntable ensured that a single swipe of a paw was not sufficient to bring the food within reach, and it took many sessions before the cats had learned that careful manipulation with one paw was needed to pull the food within reach. However, two cats that had never used the apparatus themselves, but had watched their sisters successfully rotate the turntable and eat the food, both solved the problem in less than a minute. It was probably no coincidence that it was sisters, not unrelated cats, that found it easy to learn from one another. Transmission of skills between family members may be mutually beneficial and give family groups an advantage over solitary cats, that can learn only by trial and error.

Whether young cats ever learn much from cats that are not members of their family is doubtful. Underlying distrust probably focuses their minds exclusively on staying out of trouble, overriding any curiosity about what another cat is doing. However, within a close-knit family group, younger members may benefit from watching how older,

The turntable used to test cats' ability to learn from other cats. The cat can get the food by gradually rotating the table until the bowl passes through the gap.

more experienced members of the group solve everyday problems. Because cats hunt alone, this is most likely to occur when the cats are in their shared core territory, perhaps when scavenging for food or interacting with people.

Although we may logically assume that cats began to live in family groups only since they started to associate with humans, wildcats also may have (or may have had) this ability. The formation of a group of cats seems to have only one essential requirement: a reliable source of food that can feed more than one cat and her litter of kittens. The only felid that has consistently achieved this leap is the lion, which has adopted group hunting as a way to prey on large animals. However, other felids might once have lived in small groups even without developing this additional skill.

Before man's domination of the environment in the twentieth century led to the depletion of the small cats and their favorite prey, wildcats might occasionally have lived in colonies. Several accounts left by early twentieth-century European explorers of Africa give tantalizing glimpses of this possibility. Willoughby Prescott Lowe, one of the last noted collectors of animals for the British Natural History Museum, describes a specimen he took from near Darfur in the Sudan in 1921:

> I trapped an interesting cat near Fasher. Something like a domestic cat—but v. different in coloration. The curious thing is that they live in colonies in holes in the open plain—all the holes are close together—just like a rabbit warren. I'm told they are v. local—Anyhow a cat with these habits was quite new to me! They feed on gerbils which swarm everywhere and the ground is always a mass of holes.[8]

Ten years later, on an expedition in the Ahaggar Mountains in the center of the Sahara desert, Lowe again recorded colonies of wildcats, living in burrows previously dug by fennec foxes.

Both these cats looked like typical African wildcats, but their sociability need not have come from their wildcat ancestry. The DNA of apparent wildcats from farther south in Africa, and also from the Middle East, reveals extensive interbreeding between domestic and wild cats. What Lowe saw may have been colonies of hybrids, which had retained the domestic cat's ability to live in family groups while outwardly appearing to be wildcats. That social groups of these *Felis lybica* have been recorded so infrequently suggests that, when they do occur, their social skills may have originated in previous interbreeding with domestic cats.

The switch from solitary animal to social living requires a quantum leap in communication skills. For an animal as well-armored and suspicious as a cat, a simple tiff between sisters might well escalate into a family bust-up—unless, that is, a system of signaling evolved that allowed each cat to assess the others' moods and intentions. And this, it seems, is precisely what happened.

For domestic cats, my own research has shown that the key signal is the familiar straight-up tail. In cat colonies, when two cats are working out whether to approach each other, one usually raises its tail vertically; if the other is happy to reciprocate, it usually raises its tail also, and the two will walk up to each other.[9] If the second cat does not raise its tail and the first is feeling especially bold, it may approach nevertheless, but obliquely. If the second cat then turns away, the first cat occasionally meows to attract its attention—among the very few occasions when feral cats meow. Otherwise, the first cat lowers its tail and heads off in another direction, presumably judging that the other is not in the mood to be friendly. Hesitation can be risky. My research team documented instances when a cat moving in the wrong direction, even with its tail upright, was chased off by another, usually larger, cat determined to be left alone.

Observations such as these do not prove conclusively that the upright tail is a signal; it could conceivably be something that happens when two friendly cats meet, with no meaning for either. To isolate the raised tail from everything else that a real cat might do to indicate its intentions, we cut life-size silhouettes of cats from black paper and stuck them to the baseboard of cat owners' houses. When the resident cat saw an upright-tail silhouette, it usually approached and sniffed it; when the silhouette had a horizontal tail, the cat backed away.[10]

The tail-up signal has almost certainly evolved since domestication, arising from a posture kittens use when greeting their mothers. Adults of other cat species raise their tails only when they are about to spray urine, simply for hygiene. A few individuals of *Felis lybica* in zoos do raise their tails when about to rub on their keepers' legs, but these, of course, could have some domestic cat in their ancestry. Adults of the other races of *Felis silvestris* do not raise their tails in greeting, but their kittens do hold their tails upright when approaching their mother; no one has tested whether this holds any significance for the mother, so we do not know whether kittens do this as a signal or whether it's purely incidental. Therefore, it seems most plausible that the upright tail evolved from a posture into a signal during the early stages of domestication. This would have required two changes in how cats organize their behavior: one, for adult cats to perform the kitten's raised tail posture when approaching other cats, and two, for other cats to recognize instinctively that a cat with its tail raised is not a threat.[11] Once both these changes had occurred, the posture would have evolved into a signal, one that enabled adult cats to live in close proximity to each other with less risk of quarreling.[12]

Once an exchange of tail-ups has established that both cats are happy to approach each other, one of two things occurs; which of the two seems to be influenced by both the cats' moods and the relationship they have with each other. If the cats are in the middle of doing something else, and/or one cat is significantly older or larger than the other, they usually walk up to or alongside each other. Then, keeping their tails upright, they come into physical contact and rub their heads, flanks, or tails—or a combination of all of them—on each other, before separating and walking on. Any two cats from the same group will perform this occasionally, but it is typically performed by female cats greeting males, and young cats of either sex greeting females.

The precise significance of this rubbing ritual is still unclear. The physical contact in itself may reinforce the friendship between the two participants, and thereby keep the group together—a ritual that counteracts the natural tendency of cats to regard others as rivals, not allies. The act of rubbing together also inevitably transfers scent from one cat to the other, so repeated rubbing could cause a "family odor" to build up. We know that some of the cat's carnivore relatives exchange scent via rubbing rituals: for example, badgers from the same sett create a "clan odor" by rubbing their back ends together, exchanging scent between their subcaudal glands, wax-filled pockets of skin that lie just beneath their tails.[13] Cats may not deliberately exchange odor when they rub on one another; if they were doing so, they would likely concentrate their rubs on scent-producing areas of their bodies, such as the glands at the corners of their mouths, which they use to scent-mark prominent objects in their territories—but they generally don't. As such, the rubbing ritual may be mainly tactile, a reaffirmation of trust between two animals, which by accumulation reduces the likelihood of the group splintering apart.

The other social exchange that can follow the tail-up signal is mutual licking, or allogrooming. Cats spend much time licking their coats, so it is hardly surprising that when two cats lie down side by side, they often lick each other. Moreover, they tend to groom the top of the other cat's head and between the shoulders. These are areas that the supplest of cats finds hardest to groom for itself—not impossible, of course, since cats with no grooming buddy use their wrists to wipe those areas, and then in turn lick their wrists—and all cats use this method to clean their mouths after eating.

One interpretation of allogrooming holds that it is entirely accidental: two cats sitting together groom those areas that smell the least clean, oblivious that those areas belong to another cat. However, we know that allogrooming has a profound social significance in many other animals, especially in primates, in which it has been linked with pair-bonding, the building of coalitions, and in reconciliations between family members who have recently quarreled. In cats, allogrooming likely performs the same function as mutual rubbing: cementing an amicable relationship. Consistent with this is the observation that in large groups of cats consisting of more than one family, most allogrooming takes place between relatives.[14]

Libby grooming Lucy

Some evidence shows that allogrooming does reduce conflict. In artificial colonies, such as those established by cat rescue organizations, aggression is often much less prevalent than might be expected from tensions caused by forcing unrelated animals to live together; significantly, however, allogrooming is common. Moreover, the most aggressive cats often do most of the allogrooming, implying that licking another cat may be an "apology" for a recent loss of temper. Alternatively, a cat that allows itself to be groomed does so because it remembers that it was recently attacked by the same cat, and being groomed is much more pleasant than being bitten. This latter interpretation conceives allogrooming as an alternative to aggression, so placing it in a "dominance" framework, whereby one animal controls another's activities.

Some scientists have proposed that cat societies are indeed structured according to dominance hierarchies, with larger, stronger, more experienced and more aggressive cats imposing themselves on those that are smaller, younger, or more timid. The dominance concept has long been applied to domestic dogs and their ancestor the gray wolf, but has recently been the source of much controversy. Most biologists now agree that while groups of dogs (and wolves) may sometimes appear to establish and maintain hierarchies using aggression and threat, they do so only under extreme circumstances, when their natural tendencies to form amicable relationships have been thwarted.[15]

As with dogs, the apparent formation of a dominance hierarchy in cats may be a result of external pressures. Social tensions arise when unrelated cats live together, either in one of the large outdoor colonies that form around large concentrations of food, such as fishing villages, or in a household with many unrelated cats obtained at different times from multiple sources. No hierarchies are apparent in small, one-family colonies, or indeed within family groups that form part of a larger colony.

Cat society is not as highly evolved as canid society. Domestic cat society is matriarchal: each unit begins with one female and her offspring, and if enough food is available on a regular basis, her daughters will stay with her; when they produce litters of their own, care of the kittens will be shared between them. This situation is more equitable—and probably less highly evolved—than occurs in wolf society, in which juveniles assist their parents in raising the next generation of cubs but refrain from breeding themselves that year.[16] Moreover—and in contrast to the situation in the wolf pack, which typically consists of roughly equal numbers of males and females—male cats do not help with raising kittens. In some colonies, female cats have been observed as super-affectionate toward the resident male—presumably the father of their most recent litters—possibly regarding him as a first line of defense against infanticide by other males. A typical small cat colony might thus consist of a mother, her grownup daughters, their most recent litters of kittens, and one or two tomcats.

Family groups cannot expand ad infinitum, of course, because they will inevitably outstrip their food supply. Young males start to leave their colonies at about six months of age, sometimes maintaining a shadowy existence around the fringes for a year or two, but eventually forging out on their own to look for females elsewhere—thereby incidentally preventing inbreeding. Among females, relationships become increasingly strained as they must compete for space and food. Any major event, perhaps the death of the matriarch, can trigger a breakdown in the relationships between some of the cats; aggression increases, and the once-peaceful colony may irrevocably split into two or more groups. The members of the minority group may be forced to leave the area entirely, something that will have serious consequences for them; thereafter, deprived of the resources around which the original colony formed, they are unlikely to raise many offspring to maturity. In this way, over the years the central family group will usually persist, but the more successful it becomes, the more likely that some of its female members will become outcasts. The composition of the central group is also likely to be disrupted, both by other cats trying to get access to the food and shelter that the original group are monopolizing, and also by humans, who often attempt to cull the cats as their numbers rise. As a result, cat society rarely remains stable for more than a few years at a stretch.

The benefits of cooperation, while not powerful enough to produce sophisticated social behavior, have evidently been sufficient for a limited range of social communication to have evolved: the tail-up posture, mutual rubbing, and allogrooming. In the absence of evidence to the contrary, we can reasonably assume that this change did not start until cats began their association with humankind, some 10,000 years ago. If so, it has happened remarkably quickly, but not impossibly so. Although we usually conceive of evolution by natural selection operating over time scales of hundreds of thousands or even millions of years, examples of exceptionally rapid change or "explosive speciation" have recently been documented in wild animals that have stumbled on new and hitherto unexploited environments, such that completely new species can emerge in just a few hundred generations.[17] Furthermore, if the tail-up signal has evolved from a posture performed by kittens into a sign whose meaning is recognized by every adult cat, then it is the only documented example of a new signal having evolved as a consequence of domestication; every other documented domestic species communicates using a subset of signals performed by its wild ancestor.

Male domestic cats, in contrast to females, appear largely untouched by domestication, apart from their capacity to become socialized to people when they are kittens. Each one precisely resembles Rudyard Kipling's "cat that walked by himself."[18] Unlike lions and cheetahs, male domestic cats do not form alliances with one another, remaining resolutely competitive for the whole of their lives—which, as a result, are often both eventful and rather short. Female cats (and neutered males) try to avoid one another when they can, but when two males meet, and neither wishes to back down, their fighting can be brutal (see box on page 176, "Bluff and Bluster").

Tomcat spraying

Bluff and Bluster

When two rival cats meet, each will try to persuade the other to back down without either having to resort to physical contact. Cats are too well armed to risk fighting unless this is unavoidable; they therefore resort to adopting postures that attempt to persuade their opponent that they are bigger than they really are.

Each cat will draw itself up to its full height, turn partially sideways, and make its hair stand on end, all designed to make its profile seem as big as possible. Of course, since both cats do this, neither gains any advantage, but that also means that neither can risk not performing this display to its maximum extent. The only clue that such a cat may be less than confident about winning is when it pulls its ears toward the back of its head: the ears are very vulnerable to being damaged in a fight, as testified by how ragged they can become even when they belong to the most successful of toms.

At the same time, each cat tries to add to the general effect by uttering a variety of calls, each designed to enhance the general impression that it is not to be trifled with. These include guttural yowls and snarls, violent spitting, and especially low-pitched growls—the lower the sound, the larger the voice box must be, and therefore by implication, the larger the cat. These vocalizations usually continue even if the visual posturing fails and actual fighting begins.

Cats, lacking the dog's rich repertoire of visual signals, find it difficult to signal an intention to back down. Fights usually end with one cat fleeing, with the victor in hot pursuit. If neither cat wishes to fight, one will gradually adopt a much less threatening posture, with its body crouched and its ears flattened, and then attempt to creep slowly away, frequently looking over its shoulder to check that the other cat is not about to launch an attack.

Because most owners have their cats neutered, mature tomcats are something of a rarity in western society. Few people keep them as pets, and many of those who try to do so are subsequently discouraged by the pungency of the urine tomcats spray around the garden (or, worse, the house); by the wounds they receive from stronger, more experienced tomcats in the neighborhood; and by their weeklong absences as they journey off in search of receptive females. Most owners of male cats never get to this point, taking the advice of their veterinarian to have their kitten neutered before testosterone starts to kick in at about six months of age. Male cats that have been neutered during their first year behave much more like females than males, and are usually as sociable toward other cats as a female would be under the same circumstances; that is, most will remain friendly toward other cats that they have known since birth (usually, but not necessarily, their blood relations), and a few will be even more outgoing.

Tomcat behavior gives us insight on how the cat may be evolving today. A tomcat's main goal is to compete for the attention of as many females as possible. As a consequence, wildcat toms evolved to be 15 to 40 percent heavier than females. This is also true of domestic cats today; tomcats' physicality appears to have been affected little by domestication.

By definition, half the genes in each new generation of kittens comes from their fathers. Because successful tomcats can mate with many females over the course of a lifetime, those tomcats that leave the most offspring have a disproportionate effect on the next generation. Most owners of unneutered female cats allow them to mate with any tomcat in the neighborhood, so the decision over which tomcats leave the most offspring is usually determined by the cats themselves, and not by people.

The tactics tomcats employ to maximize their chances of successfully mating are affected by how many females live nearby. Where females are widely dispersed, as with wildcats and rural domestic cats, males try to defend large territories that overlap with those of as many females as possible, typically three or four. Even allowing that tomcats are larger than females, the amount of available food in such large territories is far more than they need, but their primary goal is access to females, not nourishment. Inevitably, there are not enough females for all males to monopolize more than one female—male and female cats are born in roughly equal numbers—so some tomcats, usually the younger ones, must adopt a different strategy, that of roaming around, trying to find unclaimed females while avoiding bumping into those males that are established territory-holders.

This overtly competitive system inevitably breaks down when large colonies of cats form around an abundant source of food. Although tomcats' tactics seem to have evolved prior to domestication, when all females were solitary and lived in separate territories, there appears to have been no great need for males to change their behavior when females began to live in small family groups, attracted by the fairly modest concentrations of food stemming from human activity. This arrangement continues to this day, for example on farms that can support just a handful of cats.

In places where scores of females are concentrated into one area—fishing ports, towns with many outdoor restaurants, or where several feral cat-feeders operate together—no single male, however strong and fierce, could succeed in monopolizing several females, or even one. In these situations, there is a buildup of large colonies that include both males and females, some of which act on their own, and others cooperating in family groups. Each male is essentially in competition with all the others for the attention of the females, but, as we have seen, they somehow manage mostly to avoid the overt fighting that occurs around smaller colonies. Moreover, males in these large groups are often even less aggressive when some of the females become receptive to mating, almost as if they knew they needed to be on their best behavior for a female to accept them as a mate.

Most of the time, female cats avoid contact with males, particularly those they don't know well, presumably for fear of being attacked. There are some exceptions. In the small, single-family colony that David Macdonald studied, all the females behaved affectionately toward the resident tomcat, possibly hoping that he would defend their kittens against any marauding males attempting to take over the group. Of course, this antipathy changes as the female begins to come into season. When she is in the pro-estrus phase, the few days prior to actual mating, she becomes both more attractive to males and more tolerant of them, although at this stage she will not allow more than fleeting contact. She becomes more restless than usual, and repeatedly deposits scent by rubbing against prominent objects in her territory. If there are no males in the vicinity, she begins to roam away from her usual haunts, scent marking and uttering a characteristic guttural cry as she goes. She probably also advertises her imminent willingness to mate by changes in her scent, that may be detected by males as far away as several miles downwind. Then, as estrus approaches, she begins to roll over and over on the ground, purring all the while, interspersed with bouts of stretching, kneading the ground with her claws, and more restless scent marking. By this time, several males are usually in attendance. While she now lets them approach, she does not allow them to mount her, fighting off any that try with her claws and teeth.

As each male attempts to maximize the numbers of kittens they sire, each female tries to maximize the genetic quality of the limited number of kittens she can produce in a lifetime; she uses her courtship period to select between the males that she has attracted. She may already know something about them from the scent marks they have deposited in her territory (see box on page 180, "That Catty Smell"), but she can make a more balanced assessment from observing their behavior toward one another and toward other females. She may then decide—or be forced—to accept more than one for mating.

As her hormones take her into full estrus, the female's behavior changes abruptly. In between bouts of rolling, she crouches with her head close to the ground, treading with hind legs partially extended, and holding her tail to one side, inviting the males to copulate. The boldest of those in attendance then mounts her, grasping the scruff of her neck in his teeth. A few seconds later, in apparent contradiction

That Catty Smell

Female cats can be very choosy when it comes to selecting a father for their kittens, so tomcats need to advertise just how successful they are, preferably before they ever actually meet the female. They probably do this through the pungent smell of their urine—repellent to our noses, but presenting vital information to a cat's. To make sure that this smell is picked up by as many other cats as possible, the tom doesn't squat down before urinating, as other cats do, but instead backs toward a prominent object such as a gatepost, raises his tail, stands on the tips of his hind toes, and sprays his urine onto the object as high up as he can.

The powerful odor of a tomcat's urine, much stronger than that of a female or a neutered male, is caused by a mixture of sulfur-containing molecules called thiols, similar to those that give garlic its characteristic smell.[19] These do not appear in the urine until it has actually been voided and come into contact with the air—otherwise the tomcat himself would smell as if he ate garlic for every meal. In the bladder, they are stored in an odorless form, as an amino acid that was first discovered in cats and hence was given the name felinine. Felinine in turn is generated in the bladder by a protein called cauxin.

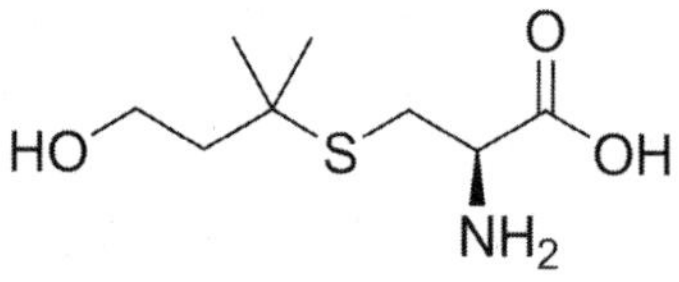

Felinine

In tomcat urine, the signal is probably not the protein (cauxin) but the felinine, which is made from one of two amino acids, cysteine and methionine, both of which contain the sulfur atom necessary for the eventual generation of the pungent odor. Cats cannot make either of these amino acids for themselves, meaning that the amount of felinine they can make is determined by the amount of high-quality protein in their diets. In a wild cat, this is in turn determined by how successful a hunter that cat is. Thus, the smellier the urine mark is, the more felinine it contains, indicating that the cat that made the mark must be good at obtaining food.

(continues)

That Catty Smell *(continued)*

The urine mark must also contain information about the identity of the cat that produced it; otherwise, a female could not know which of the several males in front of her was the best hunter. So far, scientists have not investigated this part of the message, but it is presumably other than the pungent sulfur compounds. Other species use the vomeronasal organ to detect individual odor "signatures."

However repellent to us, the cat-urine odor is a genuine badge of quality. A sickly or incompetent male is simply unable to get enough food to make its urine pungent. The evolution of this signal was therefore probably driven by the females, which selected males based on how smelly their urine was: males unable to make felinine, however good their diet, would not have been favored by females. Using these criteria, tomcats whose owners feed them high-quality commercial food are technically cheats, but this has happened too recently to have had any effect on female behavior—and anyway, I guess nobody has told them yet.

to her invitation, she screams in pain, turns on him, and drives him away, spitting and scratching. This abrupt change of mood is brought about by the pain she undoubtedly experiences during copulation: the male cat's penis is equipped with 120 to 150 sharp spines, designed to trigger her ovulation (cats, unlike humans, do not ovulate spontaneously, but require this stimulus). Happily, she appears to forget this discomfort quickly; within a few minutes, she displays herself to the males all over again—a cycle that continues at a gradually decreasing rate for a day or two. Her receptive period over, she removes herself from the area and the males disperse. If she has not become pregnant, she comes into season again every couple of weeks, until she successfully conceives.

Much of this complex ritual evolved long before domestication, when every female cat lived in her own territory, perhaps several miles away from the nearest male. Scientists think that the female's delayed ovulation evolved to give her plenty of time to locate a mate, since wildcats, unlike many carnivores, do not usually form pair bonds. This drawn-out courtship would be unnecessary if only one male was attracted at a time. Instead, it seems designed for the female first to attract several males, presuming that there are a few good ones in the area, and then to observe them for many hours or even days, so she can gauge which is likely to provide the best genes for her kittens—which is all she can hope to get from their father, since paternal care is unknown in cats.

The optimum strategy for males probably changed once man began to provide cats with food, at the beginning of the journey to domestication. Female territories became smaller and reliably focused on prime hunting areas around granaries, rich areas for scavenging, and any places where people deliberately provided food. It then became efficient for males to compete all year for territories that encompassed those of several females, and to monopolize those females when they came into season. Males that adopted the "old-fashioned" tactic of roaming widely, expecting to pick up the scent of receptive females by chance, were likely less successful. The few studies that have investigated male mating patterns show that the most successful males are often those that combine these tactics, staying "at home" when their own females were likely coming into season, but also making brief forays to other groups and solitary females nearby in the hope of achieving successful mating there also.[20] Such a no-holds-barred lifestyle takes its toll, however: males are rarely strong or experienced enough to compete effectively until they are three years old, and many do not survive beyond six or seven, the victims of road traffic accidents or infections spreading from wounds incurred in fights.

When many cats of both sexes all live in one small area, tomcats are forced to change their tactics. Monopolizing even one female risks becoming unprofitable, since while his back is turned to deal with one challenger, another male may sneak in and successfully mate with "his" female. Under these circumstances, the toms appear to accept that each female will inevitably copulate with several males; instead of attempting to attach themselves to any one or two individuals, they try to mate with as many females as possible.

This strategy probably did not evolve specifically to allow cats to live at such very high densities. Indeed, such high-density cat colonies probably occur rarely in nature, although they attract a disproportionate amount of attention from scientists where they do, given how convenient they are to study. Instead, males that grow up in such large groups probably learn which tactics work best—or, perhaps more likely, which tactics should be avoided because they risk injury—from a combination of personal experience and observation of how older males in the colony conduct themselves. They may also use this knowledge as the older males become too infirm to compete: in the few large colonies that have been studied, young males rarely emigrate, the opposite of the situation in smaller colonies. This situation may result in some inbreeding, given the enhanced risk that a male that remains in his family group will mate with a close female relative. Still, immigration by males and females from outside, attracted by abundant resources—usually a "cluster" of dedicated cat-feeders—prevents such potential inbreeding from becoming debilitating.

For some time, scientists have known that the coat colors of the kittens in some litters can be accounted for only if some had one father and some another. If a female attracts several males, she will sometimes refuse all but one; often, however, she will choose to mate with two or even more. Multiple paternity within a single litter is therefore always a possibility, but we cannot reliably detect it by coat color alone, especially in large colonies where the males may all look similar. DNA testing has enabled much more reliable examination of how females' opportunities for choices at mating translate into actual choices of fathers for her kittens.

In small colonies with one resident male, females do not appear to have exerted much choice at the time of mating: only about one litter in five exhibits DNA from any other male. The females may, however, have settled on their choice of male long beforehand, when they allowed him to live alongside them. Alternatively, he may simply have imposed himself on them, giving them little option but to wait for another, larger male to turn up if they had doubts about the quality of his genes. At present, we have little scientific evidence either way.

In larger colonies, females not only mate with several males in sequence, but the large majority of their litters also contain the DNA of more than one of these. The females may be exerting some choice—they sometimes show a preference for males from outside their group, thus preventing inbreeding, and also for those males from their own group that are able to defend the largest territories.

The female's practice of offering herself to several males one after the other may have another purpose: protecting her future offspring from infanticide. Each male has observed her being mated by others, but cannot know which of her kittens are his and which not. Likewise, since each of the larger males has mated with several females, none has any incentive to kill any litters.

Most urban males today are faced with a new and different problem: how to locate females that are capable of breeding. Nowadays, an increasing proportion of pet cats are neutered before they are old enough to reproduce. Animal welfare charities not only promote the spaying of all females before they have even one litter, they also attempt to seek out and neuter the feral colonies that remain.[21] The urban tomcat is unlikely to locate, let alone defend, a harem of reproductive females. In built-up environments, most tomcats probably adopt the same roaming lifestyle as their wild forebears, hoping to stumble on a young female whose owners have delayed spaying either because they want her to have a litter or simply because they are unaware that today's reliable nutrition enables females to mature more quickly than ever before and can come into season when they are as young as six months old.

Judging by my admittedly casual observations, tomcats seem unable to distinguish between neutered females that form the large majority in many towns and cities, and the few, mostly young, females that are still reproductively intact. Roaming toms still visited annually in late winter to check out my two females ten years after they had been neutered (having produced one and three litters respectively). Spayed females may be difficult to distinguish from intact females between seasons, certainly not by the way they behave and possibly not even by their odor, to which tomcats are presumably very sensitive.

Urban toms thus face a needle-in-a-haystack problem: they are surrounded by hundreds of females, only an otherwise indistinguishable few of which will ever present them with the opportunity to sire any kittens. The tomcats must therefore roam as widely as possible, endlessly straining their senses for the yowl and odor of the rare female that is coming into season. Such toms are shadowy animals; some are theoretically "owned"—though their owners rarely see them—and many feral. Because they make themselves inconspicuous except when they have located a prize female, there are probably far more of them than most people realize. When it first became possible to obtain a cat's DNA fingerprint from just a few hairs, my research team attempted to locate every litter born in homes in a couple of districts of Southampton, UK. From what we'd read, we expected to find that just a few "dominant" tomcats had sired most of the litters in each district; instead, we found that out of more than seventy kittens, virtually all litters had different fathers, only one of which we were able to locate. We found no evidence for littermates having different fathers, which implied that most estrous females had attracted only one male. Apparently, by inadvertently "hiding" the few reproductive queens that remain in a sea of spayed females, the widespread adoption of neutering is making it difficult for even the fiercest, strongest tomcat to do much more than search at random, thereby giving all the males in the area an even chance of reproducing.

Cats of both sexes have shown remarkable flexibility in adapting their sexual behavior to the various scenarios that we humans have, over the course of time, imposed on them. One reason that they have coped so well is simply that female domestic cats are very fecund, capable of producing as many as a dozen kittens per year. Even when conditions for reproduction are difficult, most free-living female cats will leave at least two or three descendants, enough to maintain the population even if many of her litters do not survive to adulthood.

Cat society has even adapted to life at high densities, despite origins in territorial rivalry. Logically, that heritage should result in mayhem, totally unsuitable for raising defenseless infants. Females seem to have resolved this problem by accepting the advances of several males each time they come into season, thereby making each male uncertain which kittens are his, and so banishing all thoughts of infanticide. When cat colonies are smaller, females choose to live in small family groups, and may bond with only one male, whom they accept as being both the most suitable father for their kittens and the most effective at driving away marauding rivals. When they are deprived of the company of their female relatives, either because food is insufficient for more than one cat, or due to human intervention, females are equally capable of bringing up litters without any assistance whatsoever. This remarkable flexibility must have contributed to the cat adapting to such a wide range of niches.

Male cats have also adapted remarkably well, though in a different way to females. Since in theory one male cat can sire many hundreds of kittens, the number of males in any one area is never likely to be the factor that limits the size of the cat population. Nevertheless, each individual male has no interest in the survival of his species, only in producing as many offspring of his own as he can. When breeding females are few and far between, a male must search as widely as possible for mating opportunities, pausing only to grab enough food to keep him on the move. If he can be the first and ideally only male to reach a female at the crucial moment, she is unlikely to be choosy about his quality. If he happens to be in an area where there are several females, and some of these live in groups, it may benefit him to form a pair bond or harem—although this seems not to inhibit him from searching for additional females elsewhere, when he judges that he can get away with it. He will also keep an eye out for other groups of females that he might try to take over in the future, and it is most likely males that are in this frame of mind that commit the occasional act of infanticide. However, if he finds himself living in a vast colony with many potential rivals as well as many potential mates, he will learn to curb his natural aggression, this being the only sensible strategy that will allow him to father some kittens without sustaining mortal injury.

These tales of sex and violence seems far removed from the cozy world of cat and devoted owner, but these are of course the ways in which the next generation of cats is created. Only about 15 percent of cats in the Western world come from planned mating, the great majority arising from liaisons initiated by the cats themselves. Since most pets are now neutered before they become sexually mature, most cat owners are only dimly aware of the shenanigans that undoctored cats get up to; spayed females behave as if permanently stuck between cycles of reproductive activity, and males that are neutered young never develop characteristic tomcat behavior, instead behaving more like neutered females. Tolerance of other cats is improved by neutering, but only up to a point, and family bonds established between brothers and sisters, or mothers and their offspring, are still apparent if the cats continue to live in the same house. Unrelated cats that are brought together by their owners to live in the same household, or that meet on the boundaries between adjacent gardens, still often display the natural antipathy that they have inherited from their wild ancestors. However, unlike those ancestors, today's domestic cats can also establish close bonds with humans, bringing a new dimension to their social networks.

CHAPTER 8

Cats and Their People

The relationship between cat and owner is fundamentally affectionate, surpassed in its richness and complexity only by the bond between dog and master. Cynics often suggest that cats trick people into providing food and shelter through false displays of affection, and that cat owners project their own emotions onto their cats, imagining that the love they feel for their pet is reciprocated.

We cannot dismiss these claims lightly, but surely we feel such affection for cats with good reason. Ferrets, which can be just as effective at pest control as cats, have never found a place in the hearts of the majority—although they do, of course, have their fans. Our emotional bond with cats does not stem from gratitude for mere utility; in fact, many of today's cat owners find themselves disgusted by their cat's hunting prowess, while continuing to love them as pets. So it is indeed possible that we humans are somewhat credulous, drawn in by some quality cats possess that encourages us to anthropomorphize their behavior.

The most obvious reason we might think of cats as little people is the humanlike qualities of their facial features. Their eyes face forward, like ours and unlike those of most animals—including ferrets—which point more or less sideways. Their heads are round and their foreheads are large, reminding us of a human baby's face. Infants' faces are powerful releasers of caring behavior in humans, especially in women of childbearing age. The effects of cats on humans can be remarkable: for example, scientists have found that simply viewing pictures of "cute" puppies and kittens temporarily enhances people's fine-motor dexterity, as if preparing them to care for a fragile infant.[1]

Our preference for baby-faced animals is exemplified by the "evolution" of the teddy bear. Originally naturalistic depictions of brown bears, teddy bears changed over the course of the twentieth century, gradually becoming more and more infantile: their bodies shrank, their heads—especially their foreheads—grew, and their pointed snouts were transmogrified into the button nose of a baby.[2] The "selection pressure" that caused these changes did not come from the children that played with the bears—children of four and under are equally happy with a naturalistic bear—but instead the adults, mostly women, who bought them. Cats' faces did not have to evolve to appeal to us; they have always had the right combination of features to appeal to people. That's not to say they have achieved their maximum "cuteness": freed from the constraints of biological practicality, cat images have continued to evolve, reaching their apotheosis in the Japanese cartoon figure "Hello Kitty," whose head is bigger than her body, and whose forehead is larger than the rest of her head.

Cats have a built-in visual appeal that we find attractive, but while this may persuade us that it might be pleasurable to interact with them, it cannot possibly be sufficient to sustain an affectionate relationship. Indeed, pandas look very appealing—for much the same reasons—and their image has helped raise millions of dollars for the World Wildlife Fund which adopted the panda as its logo more than fifty years ago. The WWF acknowledges the key role that the panda's apparent cuteness has played in its success, with fundraising campaigns such as "The Panda Made Me Do It."[3] Yet no one who has encountered a panda would assert that it would make a great pet. They simply don't like people—or each other, much. Thus superficial cuteness, while important, cannot be the only quality that endears us to cats.

Cats owe their success as pets not just because of their looks, but because they are open to building relationships with humankind. As they became domesticated, they evolved an ability to interact with us in a way that we find appealing, and it is this that has enabled them to make the transition from vermin killer to treasured companion. Our need for cats as pest controllers has waned, but despite this they have increased in numbers, with the majority now pets first and foremost.

We feel affection for our cats, but what do they feel for us? Wildcats generally regard people as enemies, so the answer must lie in the way cats have changed during domestication. Cats have not traveled as far down this road as dogs. Dogs have largely worked *with* man, herding, hunting, and guarding, and have therefore evolved a unique ability to pay close attention to human gestures and facial expressions. Cats have worked *independently* of man, going about their pest-controlling tasks alone and acting on their own initiative; their primary focus had to be on their surroundings, not on their owners. Historically, they have never needed to form as close an attachment to humans as dogs have. Nevertheless, even at the earliest stages of domestication, cats needed humans to protect and feed them during the times when the vermin they were supposed to eradicate were in short supply. The cats that thrived were those that were able to combine their natural hunting ability with a newfound capacity to reward people with their company.

Cats' attachment to people cannot be merely utilitarian; it must have an emotional basis. Since we now know that cats have the capacity to feel affection for other cats, why should they not feel the same emotion toward their owners? Domestication has enabled cats to extend their amicable social bonds to include not only members of their own (feline) family, but also to members of the human family that takes care of them. Because cat society is nowhere near as sophisticated as canine society, we cannot expect the same degree of devotion from every cat as dogs typically display toward their masters. The lasting loyalty feral cats can show toward members of their own family is the raw material on which evolution has worked, resulting in a new capacity to form affectionate bonds toward humans as well.

Every time I'm tempted to think that the skeptics might be right, and that cats only pretend to love us, I think back to one of my own cats, Splodge. He was a neutered, long-haired mutt-cat, as standoffish as his mother and sisters were outgoing and affectionate. He liked to sit on his own in a corner of the room, never on anyone's lap. If visitors came to the house, he would get up with an air of supreme reluctance, stretch himself, and slowly leave the room. He wasn't afraid of people; he just didn't like being disturbed. Yet, Splodge did like a couple of select people. One was a research student of mine who came to the house to play games with him as part of her doctoral studies. After the first few times, he came running as soon as he heard her voice at the door, presumably recalling the fun he'd had on her previous visit.

The other target of Splodge's affections was, thankfully, me. Whenever he noticed I had taken the car to work, he would sit in the front garden all day, even if it was raining, and wait for me to come home. On seeing my car return home, he would come running across the driveway and sit by the car. As soon as I opened the car door, he would push his way in, purring loudly. After a brief but excited tour of the car's interior, he would stand with his hind legs on the passenger seat and his front paws on my leg, and rub his face on mine. It is hard to argue that such a display was not driven by a deeply felt emotion, and by all appearances, that emotion must have been affection. This affection was unlikely motivated by a desire for food or reward—my wife usually fed Splodge, yet he never behaved in this way toward her.

The most convincing indicator that cats feel genuinely happy when they're with people was found by accident more than twenty years ago. Scientists were investigating why some wild felids were proving especially difficult to breed in captivity, guessing that many of the females were so stressed by being kept in small enclosures that they were unable to conceive.[4] To develop a method for assessing stress, the scientists relocated two pumas, four Asian leopard cats, and one Geoffroy's cat from their familiar enclosure into an unfamiliar one. These are all territorial species, so the loss of their familiar surroundings should have caused them considerable anxiety. The scientists measured how much stress hormone (cortisol) appeared in each cat's urine, and, as they expected, it increased dramatically on the first day after the move, settling gradually over the next ten days or so as the cats adjusted to their new surroundings.

By way of comparison, they also analyzed the urine of eight domestic cats that were kept in zoo-type enclosures. Four of these were known to be very affectionate toward people, whereas the other four were somewhat unfriendly. All eight were given a daily veterinary examination, something many cats find mildly stressful, and as expected, this caused an increase in the amount of cortisol excreted by the four less sociable cats, showing that they had indeed felt stressed. When in their usual environment, the level of stress hormone in the four affectionate cats' urine was slightly higher than in the other four, suggesting that they specifically did not like being caged; when the additional daily handling by the veterinary staff began, their stress levels went down. Thus, these cats appeared to be mildly upset when left alone, but contact with people—even contact that the average pet cat might object to—had a calming effect. Although it is probably going too far to suggest that these cats were suffering from "separation anxiety," they seem to have been happiest when they were receiving attention from people.

With no access to a hormone assay lab, owners can judge their cats' emotions only by their actions, and cats do not wear their hearts on their sleeves as dogs do. For most animals besides dogs, we can more logically examine the way they communicate as attempts to manipulate the behavior of other animals. Apart from animals that spend their entire lives in permanent social groups, evolution selects against total transparency. If every other member of its species is potentially a rival, it is not in an animal's best interest to shout with joy every time it finds something tasty to eat, a safe place to sleep, or an ideal mate. Conversely, if an animal is in pain or unwell, it attempts to hide all outward signs of discomfort for fear that this weakness will encourages a rival to drive it away. In solitary animals such as cats, exaggerated displays of emotion are therefore unlikely because evolution would have selected out these behaviors. Individuals that give little away leave more offspring than those whose behavior betrays their fortune, both good and bad.

Honest indicators of an animal's strength and health, such as a peacock's tail or (in my opinion) a tomcat's pungent urine, usually evolve only when the interests of one set of animals—in both these instances, the females—are distinct from those of another set—the males. For their part, females evolve a way they can discriminate between males that are successful hunters from those that are not, one that the males cannot fake. The male peacock's huge tail is the classic example, requiring its bearer to be supremely healthy; a less-than-perfect tail clearly indicates to peahens that they should look elsewhere for a mate. Likewise, any tomcat that produces urine that stinks of the breakdown products of animal protein must be a good hunter, with no room for faking.

On the other hand, using a signal is a low-cost way of getting something—a means of obtaining something from another animal without a fight. For example, when kittens are tiny, their mother has no choice but to feed them: it is the only way they can survive, and she has already invested two months of her life in them. As they get older, it is in her best interest to persuade them to take solid food, but they often try to continue to take milk from her—and one way they do this is by purring.

Here we have the first indicator that purring, the classic sign of a contented cat, might also mean something else. Purring is a rather quiet vocalization (see box on page 194, "The Rumble: How Cats Purr"), audible only over short distances, which almost certainly evolved as a signal produced by kittens, not between adult cats. In the cat's wild relatives, the only close-quarters interactions that occur, apart from mating, are those between mother and kittens. The kittens begin purring when they start to suckle, and the mother sometimes joins in. And as they start to suckle, the kittens knead their mother's belly to stimulate the flow of milk.

Cats don't purr only when they are in physical contact: some will also purr when they're walking around, trying to get their owner to feed them. This purr takes on a more urgent quality, particularly effective at getting the owner's attention. Some cats continue to purr even when their body language, for example a fluffed-up tail, says they are getting angry with their owner.[5] Not so obvious to the average owner, adult cats purr also when there are no people in the vicinity. Studies performed with remote radio microphones show that cats may purr when they're greeting another cat that they're friendly with, when they're grooming or being groomed by another cat, or simply resting in contact with another cat. Occasionally, cats have also been heard purring when they were in deep distress, perhaps following an injury or even during the moments before death.

The Rumble: How Cats Purr

Purring is an unusual sound for an animal to make; although it appears to come from the region of the voice box, it is (almost) continuous. For this reason, scientists once thought it to have been produced by the cat somehow making its blood rumble through its chest. A close examination of the sound reveals that it changes subtly between the in-breath and the out-breath, and almost stops for a moment in between. To see this in action, read the chart, known as a sonogram, from left to right: the higher the spikes, the louder the noise. The in-breaths are shorter and louder, and the out-breaths are longer.

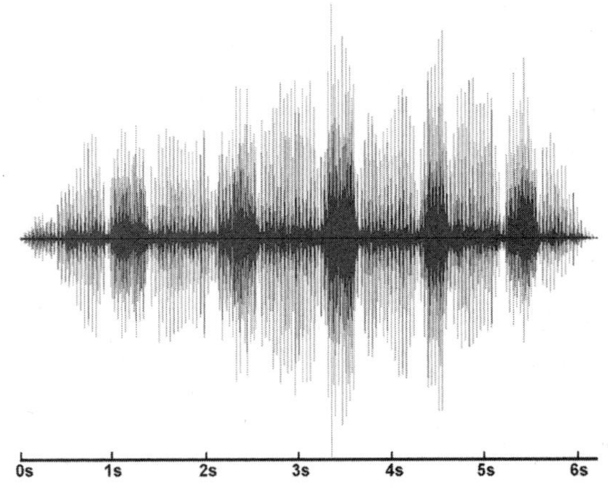

Sonagram of purring: reading from left to right, the in-breaths are shown by the dense "spikes," with quieter out-breaths in between

(continues)

The Rumble: How Cats Purr *(continued)*

The rumbling sound is produced as the vocal chords are vibrated by a special set of muscles, like a low-pitched hum. The difference from a hum is that the basic sound is not produced by air passing across the vocal cords and making them vibrate, but rather by the vocal cords themselves banging together, like an old-fashioned football rattle, ratchet, or gragger. Often the cat will also hum at the same time, but this is possible only on the out-breaths, reinforcing the sound of the purr and giving it a more rhythmic quality.

Some cats can add a further meow-like sound to their purr, making it seem more urgent to the human listener.[6] These cats often use their "urgent purr" when soliciting food from their owners, reverting to the normal purr when they are more content. This version of the purr has not yet been recorded from kittens or from ownerless cats, so it may be that, like the meow, this is something individual cats learn as an effective way to get something they want.

Purring therefore seems to convey a general request: "Please settle down next to me." In the gentlest possible way, the purring cat is asking someone else, whether cat or human, to do something for it. When they purr, kittens are asking their mother to lie still for long enough to allow them to feed—something they cannot take for granted. As in the wild, where purring evolved, there may come a time when she will feel too hungry and exhausted to stay with them, and she may have to choose between feeding them and going hunting to feed herself. Although scientists have not studied this systematically, when adult cats purr to one another, they are presumably asking the other cat to remain still. I have heard my cat Libby purring while grooming her mother, Lucy, in a rather aggressive manner, sometimes placing one paw across her mother's neck as if to hold her down.

Purring conveys information to those with an ear for it, but not necessarily the emotional state of the cat. Of course, it may occur when the cat is happy; indeed, this may be the norm. But often, a purring cat—whether a kitten suckling from its mother, or a pet enjoying being stroked by its owner—purrs not to show that it is contented, but instead to prolong the circumstances that are making it so. On other occasions, the purring cat may be hungry or mildly anxious because it is unsure how its owner or another cat is going to react; it may even be experiencing fear or pain. Under all these circumstances, the cat instinctively uses the purr to try to change the situation to its advantage.

Mark Twain lightheartedly acknowledged that a purring cat might be conning him when he observed, "I simply can't resist a cat, particularly a purring one. They are the cleanest, cunningest, and most intelligent things I know . . ."[7] To say that cats are cunning probably overstates their mental abilities: they do not deliberately and consciously deceive their owners and each other through purring. Rather, each cat simply learns that purring under certain circumstances makes its life run more smoothly.

With the purr in mind, we can see that the way our cats behave toward us can be widely misinterpreted. Science has demonstrated that a signal many owners interpret as a sign of love may sometimes mean something else. Insofar as we know, purring is not central to affectionate relationships in cat society, apart from its original role in the bond between a mother cat and her kittens. However, other signals that we tend to overlook may in fact be more genuine displays of affection. Relationships between adult cats seem to be sustained mainly through mutual licking and rubbing, so we should examine whether these also indicate affection when our cats direct such behavior toward us.

Many cats lick their owners on a regular basis, but scientists have not yet investigated why this should be. Cats that do *not* lick their owners may have been put off because their owners resisted being licked in the past; the cat's tongue is covered with backward-facing spines that work well at disentangling fur, but can feel harsh on human skin. Conceivably, some cats may lick their owners because they like the taste: some researchers speculate that they lick us for the salt on our skins, but cats don't seem to have a strong preference for salty flavors.[8] The most likely explanation is a social one, that the cat is trying to convey something to its owner about their relationship. The question is, what that something is. It may vary from one cat to another, as also seems to be the case when one cat grooms another (allogrooming).

The reason must be basically affectionate, because two cats that don't like each other never groom each other, although grooming can apparently reunite two cats that have recently quarreled. Cats licking their owners may sometimes be attempting to "apologize" for something the cat thinks it has done wrong—possibly something the owner has not even noticed, but that has some significance for the cat. However, a cat that licks its owner's hand with one paw placed over her wrist may be attempting to exert some sort of control over her. Until we know more about why cats groom one another, we can only speculate on why they groom us.

Cat owners also engage in a tactile ritual with their cats, of course, when they stroke them. Most owners stroke their cats simply because it gives them pleasure, and because the cat also seems to enjoy it, but stroking may also have symbolic meaning for the cat, possibly substituting for allogrooming in some and possibly for rubbing in others. Most cats prefer to be stroked on their heads than any other part of the body, precisely the area toward which cats direct their grooming; studies show that fewer than one in ten cats likes to be stroked on the belly or around the tail.

Many cats do not simply accept their owner's stroking passively; rather, they regularly invite their owners to stroke them, perhaps by jumping on their laps or rolling over in front of them. These rituals may not have any underlying significance, simply being mutually agreed exchanges that particular cats and their owners have learned will lead to pleasurable interactions. But while the owner has to initiate the actual stroking, most cats then indicate precisely where they wish to be stroked by offering that part of the body, or by shifting their position to place it under the owner's stroking hand.[9] By accepting our petting, cats are doing more than enjoying themselves: in their minds, they are almost certainly engaging in a social ritual that is reinforcing the bond with their owner.

Some scientists have speculated further, that the cat is also deliberately inviting its owner to take up its scent. Cats may prefer to be stroked around the cheeks and ears because those areas are equipped with skin glands that emit perfumes that appeal to other cats, and the cat wishes the owner to take on the smell of these specific glands.[10] Cats have similar glands on areas of skin that they don't generally like their owners to touch, such as at the base of the tail, so this theory implies that the cat does not want its owner to smell of these other areas. The subtle smell of the cat, virtually undetectable to our noses, will inevitably be transferred onto the owner's hand by the act of stroking, but this exchange may not have much social significance for the cat. If it did, cats would presumably be constantly sniffing our hands; while this does sometimes happen, cats do not do so obsessively. More likely, the primary function of the stroking ritual lies in its tactile component.

While touch is very important to cats, a common visual signal, the upright tail, is probably the clearest way cats show their affection for us. In the same way that an upright tail is a sign of friendly intentions between two cats, so it must be when directed at a person. When cats raise their tails to another cat, they usually wait to see whether the other will reciprocate before approaching, but this is obviously impossible when the recipient is human. Presumably each cat learns enough about their owner's body language to be able to work out, first, whether they've been noticed—they tend not to raise their tails until they have been—and second, whether the owner is ready to interact. Or at least, most cats do: my long-haired cat Splodge sometimes startled me by approaching while my back was turned, and jumping up to rub his head against the side of my knee.

Since this tail-up signal seems unique to the domestic cat, we do not know whether it evolved first as a signal to be directed at people, and then became useful to maintain amicable relationships with other cats, or vice versa. The latter seems more likely, however: because the raised tail has its origins in kitten-mother interactions, all cats are presumably born with an innate sense of its significance, and adults are therefore able to extend its use for interactions with other cats. The alternative explanation seems more far-fetched, since we would have to assume that the first people to domesticate the cat found this gesture so appealing that they deliberately favored cats that did it every time they met them.

As when two cats meet, a cat approaching its owner with tail raised will often rub on its owner's legs. The form that the rub takes seems to vary from cat to cat, and despite years of research I am still uncertain whether there is any significance in which part of the body the cat uses to rub. Some rub just with the side of their head, others continue the rub down their flank, and some routinely make contact with head, flank, and tail. Many simply walk past without making any contact at all. A few, like Splodge, jump as they initiate the rub, so that the side of the head makes contact with the owner's knee, and the flank caresses the owner's calf.

Some more nervous cats often prefer to perform their rubs on a physical object nearby, such as a chair leg or the edge of a door. Splodge's great-great niece Libby was a classic example. Even confident cats will sometimes do this when they don't know the person well, even though they are perfectly happy to rub on their owner's leg. Indeed, most times this happens, the cat is probably just redirecting its rub onto an object because it is confident that the object, unlike a familiar person, will not push it away. However, sometimes when this happens it looks as though the cat is also scent marking the object with the glands on the side of its head. Scent is certainly deposited: when I've invited such cats to rub on posts covered in paper, those pieces of paper excite a great deal of interest when presented to cats in other households. However, the redirected rubbing behavior is not performed in precisely the same way as when a cat is deliberately scent marking an object. The difference can easily be seen if the same cat is presented at head height with the blunt end of a pencil, mimicking the protruding twigs that are many cats' favorite targets for "bunting" with their heads.

Rubbing can only be a sign of affection. Because many cats rub most intensely when they are about to be fed, they have been accused of showing nothing more than what we British call cupboard love. However, few cats confine their rubbing to occasions when they are expecting to receive a tangible reward. When two cats rub on each other, they exchange no food or any other currency; after the rub, each usually continues with what it was doing before. Such an exchange of rubs is a declaration of affection between the two animals—nothing more, nothing less.

Cats also rub on animals other than cats and humans, even animals that do not understand the significance of the ritual and are unlikely to give anything in return. Splodge used to perform the tail-up/rubbing ritual on our Labrador retriever, Bruno. Bruno was already a couple of years old when Splodge arrived in our house as an eight-week-old kitten; although he had not been brought up with cats, Bruno was too laid back to think about chasing them, so Splodge regarded him as a friend from the very start. Of course, Bruno never fed Splodge—quite the opposite: as a typical Lab he was all too eager to finish up any uneaten cat food—so the cupboard love explanation cannot possibly apply. Again, the only plausible significance must be social.

Scent must become transferred from cat to owner during rubbing, but this does not seem to hold any particular significance for the cat. Most owners certainly seem oblivious to this possibility—although apparently not Mark Twain, who once observed, "That cat will write her autograph all over your leg if you let her."[11] If the primary motivation for rubbing was to leave scent behind, cats should constantly try to sniff people's legs to discover if any other cat had left its scent there. Of course, our habit of changing our clothes regularly cannot help, but the resulting confusion might lead to more sniffing rather than less. All the evidence points to rubbing, like stroking, as a primarily tactile display.

When two cats rub each other, they don't do so in equal measures, and a similar asymmetry seems to apply when cats rub on humans and other animals. When two cats of a different size approach each other, the smaller cat usually rubs on the larger, which usually doesn't reciprocate. When Splodge rubbed on Bruno, who was substantially larger, Splodge's instincts probably told him not to count on any particular response. Likewise, when our cats rub on us, our greater height, along with their knowledge that we are in control of most of the resources in the house, probably lead them to expect the rub to be one-sided. They are showing their affection for us in a way that doesn't demand a response—which is just as well, because unless we bend down and stroke them, we generally don't reply, at least not at the time.

Cats presumably find rubbing on us rewarding in its own right—if they didn't, they'd probably give up doing it—and like most of their attempts to communicate with us, they learn how to do it gradually. Kittens spontaneously rub on older cats with which they are friendly, and continue to do this as they get older. However, after they arrive in their new homes at around eight weeks of age, young cats (especially females) may take several weeks or even months to start rubbing on their new owners, as if they need time to work out how best to use this behavior to cement the relationship. Once the habit is formed, however, it seems to become fixed.[12]

Cats are more than intelligent enough to learn how to get our attention when they need to. Many use purring to persuade us to do something they want us to do, and a few invent their own personal rituals, such as jumping onto laps, or walking along the mantelpiece dangerously close to valued ornaments. However, the meow comes nearest to being their universal method of attracting our attention.

Purring is too quiet and low-pitched to be of much use for this sort of summons. Cats also have a greeting call that they use toward one another, a brief, soft "chirrup" sound; for example, mothers use this when returning to their kittens.[13] Some cats will also use this chirrup to greet their owners: my cat Splodge used to greet me with this sound when he came back from a roam around the garden. Knowing a bit about cat behavior, I would try to chirrup back at him—something he evidently appreciated, as this exchange became something of a ritual between us.

The meow is part of the cat's natural repertoire, but they rarely use it to communicate with other cats, and its meaning in cat society is somewhat obscure. Feral cats occasionally meow when one is following another, perhaps to get the cat in front to stop and participate in a friendly exchange of rubs. However, feral cats are generally rather silent animals, nowhere near as vocal as their domestic counterparts. While all cats are apparently born knowing how to meow, each has to learn how to use it to communicate most effectively.

The meow is much the same wherever the cat happens to live, confirming that it is instinctive. Every human language has a linguistic representation of this type of call:

> The English cat "mews," the Indian cat "myaus," the Chinese cat says "mio," the Arabian cat "naoua" and the Egyptian cat "mau." To illustrate how difficult it is to interpret the cat's language, her "mew" is spelled in thirty-one different ways [in English alone], five examples being maeow, me-ow, mieaou, mouw, and murr-raow.[14]

Domestication appears to have subtly modified the sound of the cat's meow. All *Felis silvestris* wildcats can make a meow sound, wherever they live, whether the north of Scotland or South Africa. Meows performed by southern African wildcats, *Felis silvestris cafra*, are lower-pitched and more drawn out than typical domestic cat meows. When researchers played recordings of these wildcat meows to cat owners, the owners rated them significantly less pleasant than domestic cat meows.[15] During the course of domestication, humans may have selected for a cat with a meow that is easier on our ears. However, it's equally possible that the meow of the direct ancestor of the domestic cat, *Felis (silvestris) lybica*, was different from that of its southern African cousin.

Feral cat meows are not as guttural as wildcat meows, but not as sweet-sounding as those of pet cats. Keep in mind that feral cats are genetically almost identical to pet cats, which suggests that their calls—as much else in cat behavior—are profoundly affected by each individual's early experiences of people. Pet kittens don't start meowing until after they're weaned, so as they become old enough to meow, they likely try out a range of meows on their owners, quickly finding that higher-pitched versions produce a more positive reaction. As with so much of cat behavior, differences between the wild meow and the domestic meow seem to be partly genetic and partly learned; domestication has enhanced cats' ability to learn how to use their meow, but may have also altered its basic sound.

Cats can also modify their meows to suit different situations: some are coaxing, and others more urgent and demanding. They do this by altering their pitch and duration, or by combining the meow with another of their calls, perhaps a chirrup or a growl. Owners often say that they know what their cats want from the tone of their meow. However, when scientists recorded meows from twelve cats and then asked owners to guess the circumstances under which the meow had been uttered, few guessed correctly. Angry meows had a characteristic tone, as did affectionate meows, but meows requesting food, asking for a door to be opened, and appealing for help were not identifiable as such, even though they made sense to each cat's owner in the context in which they were uttered.[16] Therefore, once cats have learned that their owners respond to meows, many likely develop a repertoire of different meows that, by trial and error, they learn are effective in specific circumstances. How this unfolds will depend on which meows get rewarded by the owner, through achieving what the cat wants—a bowl of food, a rub on the head, opening a door. Each cat and its owner gradually develop an individual "language" that they both understand, but that is not shared by other cats or other owners. This is, of course, a form of training; but unlike the formalities of dog training, the cat and the owner are unwittingly training each other.

If we can decode them, the meows that we inadvertently train each of our cats to use may provide us with a window into their emotional lives. Our universal recognition of the "angry meow" and the "affectionate meow" suggests that each has an underlying and invariant emotional component, as the names I've used for them imply. The cat uses the others, the "request meows," simply to attract its owner's attention. The context within which they occur provides clues on what the cat wants—whether it is sitting beside a closed door, or walking around the kitchen gazing at the cupboard where the food is kept. The meows themselves are probably emotionally neutral.

Meowing to be let indoors

Cats demonstrate great flexibility in how they communicate with us, which rather contradicts their reputation for aloofness. Cats come to realize that human beings do not always pay attention to them, and so often need to be alerted with a meow. They learn that purring has a calming effect on most of us, as it did on their mothers when they were kittens. They learn that we like to communicate our affection for them through stroking, which fortuitously mimics the grooming and rubbing rituals in which friendly cats indulge with one another. They may even learn, through our lack of reaction, that we are oblivious to the delicate odor marks that they leave behind on our furniture and even our legs.

We could consider some of this behavior manipulative, but only to the extent that two friends negotiate the details of their relationship. The underlying emotion on both sides is undoubtedly affection: cats show this in the way they communicate with their owners, using the same patterns of behavior that they employ to form and maintain close relationships with members of their own feline family.

Owners who expect long, intense interactions with their cat are frequently disappointed. Unlike most dogs, cats are not always ready to chat, often preferring to choose a moment that suits them. Cats are also nervous of any indication of a threat, however imaginary that might be; as such, many do not like being stared at, staring often being an indicator of impending aggression if it comes from another cat. The most satisfying exchanges between cat and owner are often those that the cat chooses to initiate, rather than those in which the owner approaches the cat, which may regard such uninvited advances with suspicion.[17]

Do cats think of us as surrogate mothers, equals, or even as kittens? Biologist Desmond Morris considers that at least two of these might apply, depending on the circumstances. When cats bring freshly caught prey home to "present" to their owners, Morris asserts that "although usually they look upon humans as pseudoparents, on these occasions they view them as their family—in other words, their kittens."[18] Mother cats do bring prey back to the nest, but presumably their behavior must be activated by some combination of hormones and the presence of kittens. Male cats and females without kittens don't do this, nor do female cats attempt to treat their owners as their kittens in any other way.[19] A much more likely explanation for the unwelcome "gift" deposited on the kitchen floor is that the cat has simply brought its prey home, intending to consume it at its leisure. The place where it caught the prey will almost certainly contain scent marks indicating that other cats may be nearby, so what better way of avoiding an ambush than to return to the protection of its owner's home? However, when the cat gets there, it seems to remember that while mice are good to catch, they are not nearly as tasty as commercial cat food—hence the prey is abandoned, to its owner's revulsion.

It seems implausible that cats think of human beings as their kittens, given the size difference between us. On the other hand, it is

"Presenting" a dead vole

logical to assume that they regard their owners as mother substitutes. Much of the cat's social repertoire appears to have evolved from mother-kitten communication. Our cats' behavior indicates that they take our greater size and upright stance into account when they interact with us. For example, many cats routinely jump up on to furniture to "talk" to their owners, and many others would probably like to do so, but recall that their owners haven't approved in the past. That they rub on us without necessarily expecting a rub in return, and the apparent exchange of grooming when they lick us and we stroke them, suggests that while they do not regard us as their mothers, they do acknowledge us as being in some way superior to them. Perhaps this is simply because we are physically larger than they are, so we trigger in them behavior that they would under different circumstances direct toward a bigger or more senior member of their feline family. Perhaps it is because we control their food supply, or at least (for a cat that hunts outdoors) the tastiest options available, mimicking the situation in which a few individuals can limit access to food for other cats in a large feral colony. Both of these analogies refer to situations that wildcats never encounter and that have existed only since cats began to become domesticated, so the underlying behavior has presumably only evolved during the past 10,000 years. Hence, we must assume that their relationship with us is still in flux. A definitive answer on how cats perceive us thus remains elusive; for the time being, the most likely explanation for their behavior toward us is that they think of us as part mother substitute, part superior cat.

An affectionate relationship with people is not most cats' main reason for living. Our cats' behavior shows us that they are still trying to balance their evolutionary legacy as hunters with their acquired role as companions. They form strong attachments not just to the people they live with, but also to the place where they live—the "patch" that encompasses their supply of food. Most domestic dogs, in stark contrast, bond to their owners first, other dogs second, and their physical surroundings third. This is why it is easier to take a dog on a vacation than a cat: most cats must feel uneasy when they're uprooted from their familiar surroundings, and certainly behave as if they do. They generally prefer to be left at home when their owner goes away.

Considered logically, well-fed neutered cats should not feel the need for a territory of their own, neither for sexual purposes nor for nutrition. Most cats that have gotten into a daily routine of being fed high-quality food by their owners do not hunt. Those that do hunt are not particularly enthusiastic—after all, they don't need the nourishment—nor do they usually consume the prey they catch, which is generally less tasty than commercial food. Nevertheless, most do still patrol an area around their homes, if their owners allow them to do so. In urban areas, many do not stray far—in one study, some only ventured about twenty-five feet from the cat-flap, and none more than fifty-five yards. In rural areas, this increased to between roughly twenty and one hundred yards, depending on the cat.[20] But what motivates these cats to pursue what appears to be an unnecessary remnant of their ancestral territorial behavior?

Regular food certainly seems to reduce the area cats patrol. Socialized cats without a permanent owner, who cannot rely on a regular food supply, travel significantly farther from their "home" base, perhaps a couple of hundred yards. This is still nowhere near as far as unsocialized feral cats travel, even those that have been neutered and are therefore no longer roaming in search of mates; these cats may range for a mile or more. Neutering of males once they become adults does not substantially reduce the distance they travel, as if they are still in the habit of locating as many receptive females as possible, even though they no longer have the ability to do anything if they did find one. It is clear that any cat that realizes it cannot rely on a regular source of food instinctively increases the size of its range to compensate.

That the ranges of well-fed, owned cats are so small suggests that many are not deliberately hunting at all. If the opportunity presents itself, they may grasp it; but without hunger goading them on, they may not do so with single-mindedness. Nevertheless, the cat's brain does not link hunting and hunger together tightly, and for good evolutionary reasons. A single mouse provides few calories, so a wildcat must kill and eat several each day. If cats waited to start each hunting trip until they felt hungry, they'd unlikely get enough to eat. So, even a well-fed cat, seeing a mouse within catching distance, is likely to grasp the moment—or, rather, the mouse.

Since most pet cats do not hunt seriously, it seems curious that they spend so much time outdoors, apparently sitting and doing nothing or meandering between the same locations they visited the previous day. When my cat Splodge was about eighteen months old, I borrowed a lightweight radio-tracking transmitter and fixed it to an elasticated "safety" collar so I could locate him wherever he went.[21] I already knew that he spent about a third of his time in our garden or on the roof of a nearby garage. Once the collar was on, I learned that he crossed the garden behind the apartment block next door into a strip of woodland beyond, about an acre in area. An older male cat lived in the apartment complex, accounting for Splodge's disinclination to remain there for more time than absolutely necessary: I'd already seen the two cats engaged in standoffs on several occasions, so they were undoubtedly aware of each other. Splodge rarely ventured beyond the trees—to my great relief, since there was a busy road not far beyond. He would sometimes remain in the same location for hours at a time, usually one of a few favored vantage points such as a branch of a fallen tree, before moving on to another site or returning home. He rarely seemed to be hunting: occasionally, he caught a mouse or a young rat, but would let birds fly past him without batting an eyelid. I often wondered, and still do, what was going through his mind as he maintained his surveillance of the same small area, day after day, year after year.

Even the best-fed cats seem inclined to pursue this outdated territorial agenda. They clearly feel the need to maintain space of their own, and many are prepared to fight other cats to hang onto it, even though they no longer need to hunt to survive. Their instinct to hunt is dampened, though not entirely eliminated, by their never going hungry. Both sides of this behavior are understandable from an evolutionary perspective. Hunting puts cats at risk, so they possess a mechanism that dampens the need to hunt actively when they have not gone hungry for a long time. However, few of today's pet cats are more than a small number of generations away from feral cats that have had to live on their own resources, and for whom a productive territory was a necessity. Going back a few generations further, when commercial cat food was neither universally available nor nutritionally complete, virtually every cat would have had to hunt for much of its food, and would therefore have to defend an area in which it had exclusive access to prey. Too few generations have elapsed for this instinctive need to have disappeared—however archaic it may seem to a pet cat's owner.

Each cat's desire to establish and then defend a territory inevitably brings it into conflict with other cats. In the countryside, pet cats live mostly in clusters, dictated by our habit of living in villages rather than evenly spread across the landscape—which would presumably be their preference, since that would match their own ancestral pattern. In this situation, they reduce conflict with other cats by each foraging out in different directions, the pattern of territories resembling the petals of a flower with the village or hamlet at its center. The somewhat relaxed notions of "ownership" of cats in rural areas, with farm cats becoming pets and vice versa, also mean that if two cats find themselves living too close together for comfort, one can usually find a vacant niche nearby.

Most cat owners nowadays expect their cat to live wherever the owners choose. Perhaps not fully understanding the cat's need to form an attachment to its physical environment, owners assume that it is enough to provide food, shelter, and human company, and that if they do, the cat will have no reason not to stay put. In reality, many cats adopt a second "owner," and sometimes migrate permanently.[22]

Surveys I have performed in the UK confirm this picture: many cats stray and get "lost" even from apparently high-quality homes. We see some clues as to what happens to many of these cats from the significant proportion of owners—as many as one-quarter in some areas—who, when asked where they obtained their cat, reply, "He just turned up one day." These were not feral cats: that they were so keen to adopt a new home shows that they had recently been someone else's pet, and were desperate to find a new owner. A few of these cats may have been genuinely lost, still searching unsuccessfully for their original home, but most were probably migrants, looking for a better place to live than their original owner could provide. For the few instances of this that I could trace back to the original home, the cat had evidently left not because it wasn't being fed properly, or was unloved; something else must have gone seriously wrong for the cat to abandon the certainties of a regular food supply and, in most cases, feelings of attachment towards its original owner. The most likely explanation is that these cats could not establish an area where they could feel relaxed, safe from challenges from other cats.

The threat could have come from the next-door cat, or even from another cat living in the same household. Within the home, opportunities for conflict are rife. Just because two cats have the same owner does not mean they are going to get along. Many pay heed to the cardinal rule of cat society: proceed with caution when meeting any cat that has not been a part of your (cat) family for as long as you can remember. Many cat owners seem oblivious to this principle, blithely assuming that when they obtain a second cat, the two will quickly become friends. Although this is generally true of dogs, cats are more likely to merely tolerate one another (see box on page 212, "Signs that Cats in a Household Do or Don't Get Along with One Another"). To reduce conflict, they often set up separate, if overlapping, territories within the house, but may continue to scrap with each other sporadically. In surveys of owners with two cats, roughly one-third report that their cats always avoid each other if they can, and about a quarter say that they fight occasionally. The two cats will probably come to respect the others' favorite places to rest—the larger or original cat will generally take the prime spots—but tension may remain if both cats are fed in the same room or if there is only one litter tray. They may also compete for the cat-flap, if one cat claims it as being within its core territory. Feeding each cat in a different room and providing several litter trays in different locations (not the rooms used for feeding) can often make the situation more tolerable for both cats.

For any cat that is allowed outdoors, other cats in the neighborhood will also be a source of conflict and stress. When a cat comes into a new household—whether or not it was previously cat-free, it will most likely find that it has been parachuted into the middle of a mosaic of existing cat territories. If there are already cats on two, three, or even all four sides, the owner's garden will almost certainly "belong" to one or more other cats; for cats, garden walls are highways, not boundaries to be respected. The new cat will have to establish its right to roam in "its own" garden, and it will be able to do this

Signs that Cats in a Household Do or Don't Get Along with One Another[23]

Cats that see themselves as part of the same social group generally

- hold their tails upright when they see one another
- rub on one another, either when walking past or alongside one another
- regularly sleep in contact with one another
- play gentle, "mock-fighting" games with one another
- share their toys

Cats that have set up separate territories within the house will tend to

- chase or run away from one another
- hiss or spit when they meet
- avoid contact with one another: one cat may always leave the room when another enters
- sleep in widely separated places; often, one will sleep high up, perhaps on a shelf, to avoid another
- sleep defensively; the cat has its eyes closed and looks as if it is asleep, but its posture is tense and its ears may twitch
- apparently restrict one another's movements on purpose—for example, one cat sitting for hours by the cat-flap, or at the top of the stairs
- watch one another intently
- look unusually tense when they're in the same room
- interact separately with their owner—for example, they may sit either side of the owner to avoid physical contact with one another

only by standing up to the other cats that previously had undisputed use of the space. Challenges over territorial boundaries may continue for years: in one survey, two-thirds of owners reported that their cat actively avoided contact with other cats in the neighborhood—and, frankly, the other third probably hadn't looked out of their windows often enough. One-third reported that they had witnessed actual fights between their own cat and their neighbor's.

Surprisingly few owners seem particularly bothered about such conflicts until they begin to affect their cat's health: a bite can turn into an abscess requiring veterinary treatment. More ominously, the cat may become so stressed, or its movements so restricted, that it begins to urinate or defecate in the house (see box below, "Signs that a Cat Is Failing to Establish a Territory Outside Its Owner's House"). Even if the various cats in an area eventually arrive at a truce, owners may inadvertently reignite the conflict by temporarily removing their cat, perhaps boarding it elsewhere for a couple of weeks while they go on holiday. Encouraged by the signs that the cat may have left for good, such as fading scent-marks, absence of sightings, one or more of the neighbors' cats may start to encroach into what was previously that cat's territory. When that cat returns, it may have to reestablish its rights all over again.

Signs that a Cat Is Failing to Establish a Territory Outside Its Owner's House[24]

- Not leaving the house even when encouraged to do so
- Waiting to be let out by the owner rather than using its cat-flap (because there might be a rival cat ready to ambush it on the other side)
- Neighbor's cats entering the house through the cat-flap
- Leaving the house only if the owner is in the garden
- Excessive time watching out of windows
- Running away from windows and hiding when another cat is spotted in the garden
- Running into the house and immediately to a place of safety, far from the access point
- Tense interaction with the owner, including rough play
- Urinating and defecating in the house, by a cat that usually goes outside to do this but feels too insecure to do so
- Spraying urine (scent-marking) in the house, especially if near to access points such as doors and cat-flaps (more likely in male cats than females)
- Other signs of psychological stress, such as excessive grooming

Increasingly, cat owners are avoiding such problems by keeping their cats indoors, although their motivation may be more to protect their pet from traffic, disease, or potential thieves (especially if it is a valuable pedigree animal) than from social stress. Restricting the cat to one relatively small area for its entire life can induce stresses of its own. Although the practice of keeping cats indoors has been common among apartment-dwellers for more than thirty years, we have little systematic research into whether domestic cats find this confinement stressful. To see how we should expect indoor cats to behave if they were stressed by confinement, we must therefore look further afield, to their wild ancestors.

Wild felids often react badly to being confined. Both "big cats" such as lions and "small cats," such as jungle and leopard cats, are prone to the habit, once commonplace in zoos all over the world, of pacing to and fro in their cages.[25] Of other types of animal, only bears are as badly affected, and like most of the cat family, they are also solitary territorial carnivores. We do not fully understand why these animals pace, but the reasons probably arise from a mixture of frustration at not having access to a large enough hunting territory, even though their nutritional needs are being more than satisfied, and "boredom": well-fed carnivores do instinctively sleep for much of the time, but many seem to crave mental stimulation when they are awake. Changing the way the wild cat is fed can provide the latter: rather than simply providing one daily meal that can be wolfed down in seconds, zookeepers now provide food several times each day, and the cats must make an effort to get at least some of it—for example, by feeding bones that have to be cracked open, or placing the food in puzzle feeders that the cat has to work at for an extended period of time.

When considering whether domestic cats kept indoors are likely to suffer, we might first examine whether they show signs of objecting to being spatially restricted, and whether they show signs of "boredom." Repetitive pacing is surprisingly rare in domestic cats, considering that this is the most common abnormal behavior observed in cats kept in zoos. This difference may have already evolved prior to domestication; when in captivity, the domestic cat's wild ancestor (*Felis silvestris*) is more prone to "apathetic resting" (taking no notice of its surroundings) than to pacing. The difference could also be a consequence of domestication. Whichever the culprit, domestic cats seem to have lost much of their ancestors' spontaneous "drive" to roam. We do not know why this loss would have benefited cats: for most of their 10,000-year history, they have had to hunt for their living.

Domestication may have given the cat a much greater flexibility in its territorial behavior. Wild felids generally feed on prey dispersed over the landscape around them, and therefore have always needed a large territory: they may need to venture farther afield if food is scarce, but in all their evolutionary history, they would never have encountered a situation where food was so locally plentiful that they could afford not to go out foraging day after day. Domestic cats, by contrast, have adapted to hunting in and around what are, by comparison, very small areas—human settlements—while retaining the ability to expand their hunting ranges quickly if food becomes scarce, provided that other cats don't prevent them from doing so.

Feral domestic cats can thus have territories 10,000 times larger than those of some owned cats that are allowed twenty-four-hour outdoor access. However, just because the species as a whole shows this flexibility does not mean that individual cats are so adaptable. Much depends on their previous lifestyle and what expectations they have acquired. A feral cat accustomed to hunting across fifty acres and then suddenly confined to a pen will be almost as distressed as its wildcat counterpart. A pet cat that has never had to hunt to survive would most likely perish if abandoned somewhere remote.

Domestic cats have probably become so flexible in their demands for space that they can, under the right circumstances, adapt adequately to indoor living. Very few cats that are allowed outdoors voluntarily restrict themselves to an area as small as even the most spacious apartment, and those that do would probably venture farther away if they were not anxious about meeting other cats. However, the additional restriction does not appear to cause undue stress. Cats that have grown up wandering wherever they choose will almost certainly be stressed by being suddenly confined indoors, even when this becomes necessary to protect the cat's health. Cats that are destined for a life indoors should probably never be allowed outdoors, so that they can't miss what they've never had.

Space that is restricted needs to be quality space. It's unlikely that wild cats value open space for the simple joy of the view into the distance; rather, they presumably gain satisfaction from how many places they can see that might be concealing their next meal. Zookeepers have tried giving big cats access to wide open spaces, but this usually had no effect on their habitual pacing, and in fact they rarely visit the additional territory. However, making the same amount of space more interesting was a much more successful strategy; zoos adapted the enclosures so that the cat could not see the entire area from any one place, requiring the cat to move around.

An indoor cat must also be kept busy, since it cannot experience the variety automatically provided by the outdoors—nor, admittedly, the anxiety of being ambushed by another cat. For the owner, this requires extra effort (see box on page 217, "Keeping an Indoor Cat Happy"), which must be balanced against the relative ease of allowing a cat to seek much of its mental stimulation outdoors. In particular, owners should allow the cat to perform as much of its "natural" behavior as possible. Although there is no specific scientific evidence to support such a recommendation for the domestic cat, it is one of the guiding principles of animal welfare that have been established for vertebrate animals in general.

The owner can provide the cat with social behavior either by spending time with it or by keeping two compatible cats together. The easiest way to achieve the latter is probably to obtain two kittens from the same litter, although even this is not guaranteed to be successful—sibling rivalry is not unheard of among cats. Hunting behavior can be simulated by giving the cat a view of some "natural" space, through "play" with prey-sized toys (since cats react to these as if they really are prey), and feeding dry cat food in a device that requires the cat to perform predatory behavior to release each piece (simulation of hunting behavior has been successfully used to restore normal behavior in captive wildcats, so it should also benefit indoor domestic cats).[26]

Keeping an Indoor Cat Happy[27]

- Allow the cat as much space to roam around as possible.
- Site the litter tray in a secluded place, away from windows from which it can see other cats.
- If possible, provide an enclosed outdoor area for the cat, perhaps on a balcony. While we have no evidence that cats need "fresh air" as such, the sights, sounds, and smells of the outdoors keep it interested.
- Indoors, provide two beds of different types. One should be on the floor, with a roof and three walls; cats are often happy with a cardboard carton placed on its side. The other should be placed up high, near the ceiling but easily accessible, with a good view of the entrance to the house or out of a window. Not all cats will use both, but most will feel safe in one or the other.
- Provide at least one scratching post.
- Play with the cat several times a day. Games with prey-like toys may satisfy the cat's urge to hunt, especially if it has been viewing birds through the window. Change the toys often to sustain the cat's interest.
- Try using a puzzle feeder containing a small amount of dry cat food. A plastic drink bottle with a few holes of appropriate size cut in the sides will keep some cats busy for hours. More complex devices are available commercially.[28]

An activity-feeder made from a soda bottle

- Provide a pot of live "cat grass." Many cats like to chew on these oat seedlings, *Avena sativa*, although why they do it is somewhat obscure.
- Don't overfeed; indoor cats are at greater risk of obesity than outdoor cats.
- If you don't yet have a cat, consider getting two littermates: they will be good company for each other.
- If you already have one indoor cat, plan ahead before getting another cat for "company." Cats that have never met before are unlikely to adapt spontaneously to sharing a confined space.

None of these may be entirely satisfactory substitutes, but any stress that the cat feels at being restricted or not allowed to carry out all its natural behavior may be counterbalanced by its being removed from the stress of being "bullied" by other cats in the neighborhood.

Considering how little time has elapsed since they were first domesticated, cats demonstrate remarkable flexibility when it comes to the amount of space they need. However, we must be careful not to provide them with too little space, or space that contains incipient threats. Pet cats have no practical need to maintain a hunting territory any more, but it is too soon for evolution to have removed the desire to do so. Unfortunately for the cats, this perceived need to roam can bring them into daily conflict with other cats, each with the same agenda. Cats, having a relatively unsophisticated repertoire for communicating with other cats, take time to "negotiate" their territory boundaries, time that increasingly we no longer provide them with.

Cats face great pressure to change their ways—and not only to adapt to modern urban lifestyles. The conservation lobby, from Australasia to the United States to Great Britain, increasingly objects to their maintaining any kind of hunting territory. To change, cats must evolve new ways to organize their behavior, at an unprecedented speed. Evolution requires variation: cats must differ from one another in the way they perceive and react to their environment, both social and physical. We still see great variation among cats in their personalities, and somewhere among these, we might find the combination of characteristics for the ideal twenty-first-century cat.

CHAPTER 9

Cats as Individuals

Cats have much in common with one another; as a species, they are highly distinctive, so what is true of one cat is also likely true of another. But cats are also unquestionably individuals, both in appearance and—more significantly for their relationship with their owner, and for their future—in the way they behave. Even scientists now talk freely of cats having their own personalities. The existence of many different personality types among today's cats gives us hope that they, as a species, have the potential to adapt to the demands of the twenty-first century and beyond. Somewhere hidden among the cats that live around us today are the genes that will enable their offspring to evolve into a slightly different kind of cat; for example, one that is better adapted to living indoors.

Of course, genes alone cannot drive such changes; the environment in which a cat finds itself plays a powerful role in guiding the development of its personality. Furthermore, cats do not have to make these changes on their own; we as owners have a wealth of strategies we can employ to help them to lead happier lives. If cats' genes were as unvarying as those of some pedigree dog breeds—many of which contain little more variation in total than does the average human extended family—then no amount of breeding for temperament would achieve much: the only way forward would be for cat owners to change the way they relate to their cats. But knowing that cats are so genetically variable, even today, provides us with two complementary ways to help them adapt to our world.

The crucial question is, how much influence do genes have? Much of the cat's personality depends on other factors. For example, whether a cat will tolerate people depends on whether it has had contact—and the right kind of contact—with people during the first eight weeks of its life. Cats that do receive such contact nevertheless vary greatly in how friendly they are to people in general, and even toward their owner. How much of this variation, not fully explained by the basic process of socialization, is inherited? Is each cat's ability to tolerate other cats due simply to whether it grows up with other cats, or are some born to be more adaptable than others?

Decoding the inheritance of personality is nowhere near as simple as the inheritance of the color or length of a cat's coat. We can track most of those visible differences among cats to twenty or so well-defined genes that operate in a highly predictable manner. If a cat's parents both have black coats, then the cat will also be black; this is not affected by whether it is born in a hedge or in a kitchen.

Genes and environment can interact in complex ways, however. Even coat colors can be affected by the environment: for example, the darker "points" on a Siamese cat's face, paws, and ears come from a temperature-sensitive mutation that prevents the hairs from taking up their usual color at normal body temperature. As newborns, these cats are whitish all over because their mother's womb is uniformly warm. As they grow and the extremities of the body become cooler, the hair there grows darker, producing the characteristic "pointed" coat. Finally, as the cat enters old age and the circulation of blood in its skin deteriorates slightly, it gradually turns brown all over.

The relationship between genetics and environment is evident in personality as well. Cat personality is influenced by hundreds of genes and a lifetime of experience, interacting together to produce the cats we see today.

To search for evidence that personality can be inherited, we might start with pedigree cats. Unlike dogs, which have been bred for different functions for many centuries, pedigree cats have been bred mainly for their looks. Deliberate selection is probably not to blame for any consistent differences in behavior between different cat breeds; we cannot expect to find differences as great as those between, say, a border collie and a Labrador retriever. However, because all pedigree cats are raised by breeders, and, at least within each country, in much the same way, any consistent behavioral differences between them are likely due to genetics.

The breeding of pedigree or "show" cats is regulated by standards laid down by the individual "breed clubs," and the best cats from each breed compete in cat shows run by organizations such as the Cat Fanciers' Association and the International Cat Association in the United States, and the Governing Council of the Cat Fancy in the UK. Well-known breeds or groups of breeds include the Persian or "Exotic" breeds, stocky cats with long hair and flat faces; the "Foreign" breeds, fine-boned, long-limbed cats such as the Siamese, Burmese, and Abyssinian; and the Domestic breeds, which as their name implies were originally derived from ordinary domestic cats local to the British Isles. Some individual breeds can be defined by a single mutation, such as the short, wavy coat of the Cornish Rex, the downy hair of the Sphynx, and the short tail of the Manx.

Many of the newer breeds of cat are simply color variations on existing breeds. For example, the Havana Brown is genetically indistinguishable from the Siamese, except that it lacks the mutation that causes most of the Siamese's coat to stay cream-colored. Some of the longer-established breeds claim ancient ancestry—for example, the Siamese breed is apparently described in the "Cat-Book Poems" written in the ancient Siamese city of Ayutthaya sometime between 1350 and 1750—but their DNA shows that the breeds have become separate entities only during the past 150 years or so.[1]

This recent evidence separates the breeds into roughly six groups, each seemingly derived from—or possibly allowed to interbreed with—local street cats. The DNA of the Siamese, Havana Brown, Singapura, Burmese, Korat, and Birman shows not only that they are closely related, but that they are also genetically similar to the street cats of Southeast Asia from which they were undoubtedly derived. The Bobtail, a traditional Japanese breed, is genetically close to random-bred cats in Korea, China, and Singapore (and, presumably those of Japan, which were not included in the study). The Turkish Van cat, as its name implies, is related to non-pedigree cats from Turkey, as well as Italy, Israel, and Egypt. The Siberian and Norwegian Forest cats are derived from longhaired northern European random-bred cats, while the superficially similar Maine Coon finds its closest non-pedigree relatives in New York state. Most of the stockier breeds—the American and British Shorthairs, the Chartreux, the Russian Blue, and surprisingly the Persian and Exotic breeds, are all closely related and are presumably derived from Western European stock. The modern Persian, even if some of its distant ancestors did come from the Middle East, seems to have lost most traces of its origins, possibly due to recent breeding to produce the flat (brachycephalic) face preferred by its devotees.

The various breed clubs usually describe typical personalities for their cats. For example, the Governing Council of the Cat Fancy (GCCF) describes the Ocicat, an American breed derived from Abyssinian, Siamese, and American Shorthair lines, as follows:

> Many owners remark on the almost doglike tendencies of the Breed, in that they are devoted to people, are easily trained and respond well to the voice, but retain their independence as a proper cat should, and are very intelligent. Because of their adaptability they are a joy to be with, they are not demanding in any way and seem to take life in their stride. Ocicats are reasonably vocal and do not like being left alone for long periods, but do make ideal companions for households with other pets, and are confident with children.[2]

Although such formal recognition permeates the world of cat enthusiasts, scientists have devoted little attention to investigating whether cat breeds have distinctive personalities. The line-breeding necessary to develop cats that breed "true"—that is, where the offspring look the same as their parents—has led to some behavioral abnormalities that have a genetic basis (see box below, "Fabric Eating in Pedigree Oriental Cats"). Because these are essentially pathologies, and isolated within a single breed or group of breeds, scientists do not classify these abnormalities as aspects of "personality." Turning to more universal cat behavior, Siamese and other Oriental cats are remarkably vocal; many develop so many variations on the meow that they seem to "talk" to their owners. Longhaired cats, especially Persians, have a

Fabric Eating in Pedigree Oriental Cats

Siamese, Burmese, and other Oriental cat breeds are susceptible to developing an unusual form of pica, the eating of non-nutritive substances. For reasons we do not yet understand, some house cats develop the habit of chewing unusual items, such as elastic bands and rubber gloves, but a significant proportion of pedigree Oriental cats not only chew but also eat fabrics. Their fabric of choice is usually wool, closely followed by cotton; synthetic fabrics such as nylon and polyester are less popular. Most of these cats start by chewing woolen items: many then progress to swallowing the chewed-off fabric chunks, or move on to other materials. In these cases, the cats appear to confuse fabrics with food. I have seen a Siamese cat dragging an old sock

A Siamese cat eating a piece of cloth

(continues)

Fabric Eating in Pedigree Oriental Cats *(continued)*

up to its food bowl, and then alternately taking one mouthful of one, and one mouthful of the other.

We also have yet to understand their predilection for wool over other fabrics. One theory held that these cats might have a craving for the natural lanolin in wool, but when I tested this directly, this idea did not stand up.

Because wool eating is largely restricted to a small number of closely related breeds, it must have a genetic basis. However, it does not seem to be inherited directly. When I surveyed the owners of seventy-five kittens produced by seven mothers, three of whom were fabric eaters and four of which were not, one-third of the kittens had become fabric eaters themselves—but many of these had "normal" mothers (their fathers' habits were unknown). Neither simple genetic factors, nor imitation of the mother's behavior, could explain why some had developed this problem while others had not.

Many of the fabric-eating cats did also show other types of abnormal behavior, such as biting their owners and excessive scratching. These also occur in non-pedigree cats and are often a sign of anxiety and stress. Among Oriental cats, fabric eating often starts within a few weeks of the cat being rehomed, when the cat may be feeling stressed by the change in its environment. Onset can also occur at around one year of age even without a move, when the cat is becoming sexually mature and starting to come into conflict with other cats either within the household or outside (even though they were valuable pedigree animals, few of the cats in my study were totally confined indoors).

Fabric eating may therefore start as a soothing oral behavior that these cats adopt when they feel especially stressed, rather like thumb sucking in human infants. Why they choose fabrics, and why chewing often turns into ingestion, is still unclear.

reputation for being lethargic and not terribly fond of close contact with people, perhaps because these cats overheat easily. Beyond such self-evident differences, we have little hard information on precisely how breeds differ in personality, and how those differences arise. Most of the information we have is based on surveys of experts—veterinarians or cat-show judges, for example—who tend to see most cats when they are away from their normal territories, and therefore may not always get a complete picture of their behavior.

One small-scale study conducted in Norway confirmed that Siamese and Persian cats do indeed behave in characteristic ways in their owners' homes.[3] Although the cats' personalities were recorded by the owners themselves (which in itself could have introduced some biases) rather than through direct observation, the Siamese were reported to be more contact-seeking, more vocal, and more playful and active than standard house cats. One in ten Siamese was regularly aggressive toward people, compared to one in twenty house cats and one in sixty Persians. Persians were generally less active than other cats, and apparently more tolerant of unfamiliar people and cats—although their apparent laziness might have simply made them disinclined to run away.

It's highly implausible that every single variation in cat personality will be traced to a different gene. Rather, breed characteristics seem to emerge during kittenhood as general tendencies, such as when making the choice whether to explore or to move away from novel objects or situations. In turn, these tendencies profoundly affect what each kitten learns, and thus how its behavior develops: tactics it learns are useful in one particular circumstance may become general strategies, used in many situations. The underlying processes have scarcely been investigated in pedigree cats, but in one study, researchers found that the ability of Norwegian Forest kittens to remember novel situations develops more slowly than those of other pedigree cats (Oriental breeds and Abyssinians), whose brains may develop somewhat more quickly than those of ordinary house cats.[4] Such slowing-down and speeding-up of the rates at which different parts of the brain grow might have long-term effects on cat personality. Many of the self-evident differences in behavior between dog breeds result from changes in the speed at which different areas of the brain develop: for example, Siberian Huskies display a full range of wolf-type behavior, while breeds with "baby faces," such as bulldogs, signal to one another in a similar way to wolf cubs of just a few weeks old.[5] However, scientists have not yet documented any such link for cats.

Differences among breeds provide useful insight into whether cat behavior might be influenced by genetics, and pedigree breeds are useful in this regard because each cat's parentage is documented. Popular males can sire many kittens, yet rarely even see any of them, so their influence on their offspring must be genetic. In the Norwegian study, playfulness, fearfulness, and confidence in encounters with unfamiliar people were all distinctively different between the offspring of different fathers, although some other traits, such as aggression to cats or people, were not. Because this study was small-scale and carried out in only one country, its details may not apply everywhere; still, the principle that some aspects of a cat's behavior are influenced by its father's genes seems likely to stand.[6]

Non-pedigree cats also vary greatly in their "personalities," spurring on the myth that a cat's temperament and its coat color are inextricably linked.[7] The British refer to tortoiseshell cats as "naughty torties"; likewise, blotched tabbies are "real homebodies," mackerel tabbies are "independent," and white patches on a cat's coat have a "calming effect" on the animal. It seems part of human nature to link outward appearance and inner character, and to continue to see those links even when evidence is to the contrary. Some scientists have speculated that the specific biochemistry that generates different coat colors also somehow affects the way a cat's brain works, showing a genetic effect referred to as pleiotropy, but little evidence has been found to support this idea in cats.[8]

Links between coat color and personality do occasionally occur among pedigree cats, and these do provide an opportunity for proper investigation because the family trees are available. The relatively restricted gene pool for each color within each breed does result in certain temperaments accidentally becoming associated with particular coat colors. At any one time, only a limited number of high-quality tomcats within each breed are available to produce the desired color; as a result, the temperament of the most popular of those tomcats—or at least those aspects affected by genetics—tends to become predominant within that section of the breed. For example, twenty years ago,

Scotland's British Shorthair cats with tortoiseshell, cream, and especially red (a rare, un-patterned version of orange) coats were relatively difficult to handle; scientists traced this characteristic back to one male with a particularly difficult temperament.[9] Likewise, cats with dark "points" on their paws and ears, even if not pedigree Siamese, are likely unusually vocal, because the gene that causes the points to appear is very rare in any cat without at least one Siamese in its recent ancestry.

Coat color and some aspect of personality can also become linked if the gene that controls the color and a gene affecting the way the brain develops happen to occur very close together on the same chromosome. Because genes are grouped together on chromosomes—cats have thirty-eight: eighteen pairs, plus two sex chromosomes—not all combinations are passed on randomly from one generation to the next. If two genes occur on different chromosomes, then the chances that a kitten will receive any particular combination of the two are essentially random. However, two genes that occur on the same chromosome tend to be inherited together. This is not inevitable, because matching pairs of chromosomes do occasionally swap sections between each other, by a mechanism known as crossing over; if the swap happens in between the two genes in question, they are then inherited separately. Such exchanges rarely occur between genes that are sited close together on the same chromosome. For example, the gene that causes a white coat ("dominant white," that is; different from albino) is situated on the same chromosome and close to another gene that causes both the eyes to be blue and the cat to be deaf, a rare example of one gene affecting both appearance and (indirectly) behavior. Blue-eyed, white cats are thus almost invariably deaf.[10] In the case of the ginger cats in rural France, the gene that suits those cats to the feral lifestyle might simply be very close to the O(range)-gene (on the X-chromosome), rather than being a direct effect of the cat being orange.

Making assumptions about a cat's personality based solely on its appearance is often misleading, but cats undoubtedly do behave in individual ways, irrespective of their color. Until about twenty years ago, most scientists considered that only humans could have "personalities," yet now this concept is widely applied to animals—and not just to domestic animals. Even wild animals behave in consistently different ways that reflect different ways of reacting to the world around them: over the past few years, the concept of "personality" has been applied to animals as diverse as lizards, crickets, bees, chimpanzees, and geese. Some individuals may be particularly bold, and therefore the first to exploit a new food source, whereas others are particularly shy and therefore less likely to run headlong into dangerous situations. The success of each strategy is likely to vary depending on what the environment is like, and as that changes, so sometimes bold individuals will do best, other times they will be the ones who perish first. In this way, the genes that influence both types persist in the species.

Some of the most complex effects of personality occur in social situations. Sticklebacks, fish that sometimes swim in shoals, can be classed as either bold or shy. When a fish has a choice of shoals consisting entirely of bold or shy individuals, it will choose to join the bold shoal, irrespective of whether it is bold or shy itself. Bold shoals usually find more food, and a shy fish will find the middle of a shoal of bold fish a good place to hide. However, to keep up with the bold shoal, it must swim faster than usual, so it temporarily starts behaving more like a bold fish. Although we don't yet know much about social effects on cat personality, such observations raise the fascinating possibility that each cat may be able to adjust its personality to fit in with those of the other animals—human, feline, and canine—in the household in which it finds itself.

We have two broad approaches to studying cat personality: watching the cats, or asking their owners. Because owners are likely to be biased, observing the cat's behavior is the only way an impression of its personality can be gained. For this reason, most of my own studies have involved recording cats' behavior. To ensure that outdoor cats would be home, I chose to observe them just before and just after their usual feeding time.[11] Since many cats interact most intensely with their owners when expecting food, and since hunger usually affects the way they interact, this arrangement had the additional advantage that all the cats would have been hungry when the observation started, and sated at the end.

While their food was being prepared, the thirty-six cats in the study acted as cats usually do when expecting to be fed: walking around the kitchen with their tails upright, meowing, and rubbing on their owners' legs. After the meal, some went straight outdoors, while others sat and groomed themselves; some interacted with their owners again, while others investigated the unfamiliar human in the room—that is, the person making the observations. So far, so obvious; but the first objective of our study was to find out whether each cat behaved in a characteristic way every time. We repeated these visits once a week for eight weeks, and found that indeed they were indeed fairly consistent—so what we had measured was probably a reflection of the cat's personality, or at least its "personal style."

A bold cat studies a scientist

From their behavior before they were fed, we separated the cats into various types. Some always rubbed around their owner's legs, purring all the while; others never did. Some walked around the kitchen much more than others, and some continually tried to attract their owner's attention by meowing; their owners didn't seem to find this particularly endearing—they stroked the quiet ones more often. Only about half of the cats took any of these tendencies to extremes; the others split their time between rubbing and meowing, and were moderately active—unsurprising given that all these traits are typical of cats.

After they were fed, some of the younger cats went straight outdoors—possibly more of a habit than a personality trait. Most, however, stayed in the kitchen for a few minutes. Several of the cats that had been more active than the others before the meal continued to interact vigorously with their owners, walking around with their tails up and meowing. Those that had paid most attention to the unfamiliar person before the meal continued to do so.

We rounded out our observations by talking to the cats' owners. Some of the differences between cats we observed seemed to reflect their personalities, but we did watch them only in one (convenient) situation. Would we have seen other sides to each cat's character if we'd also watched it when it was out exploring, socializing with (or avoiding) other cats, or curled up while its owner watched television? Several studies have examined differences in cats' reactions to people by asking owners or other caregivers to report on their cats' behavior. Inevitably, these surveys provide little information on how the cat behaves when it's alone, and owners with only one cat won't necessarily know how well it gets along with other cats.

Three aspects or dimensions of cat personality have emerged from such studies, despite the studies' limitations.[12] The first is whether or not the cat gets along well with other cats in the same household or other social group; some cats seem to be more outgoing toward other cats than others—at least, ones that they know well. Second is how sociable the cat is toward people in the household; some cats seem to value close contact with their owners more than others.

Third, and possibly most fundamental, is how bold and active or steady and cautious the cat is in general. An individual cat may possess any one of the eight possible combinations of these three basic traits: for example, one cat could be shy and retiring, but affectionate toward her owner and the other cats in the house; another might be highly active and equally affectionate toward his owner, but might keep his distance from another cat in his household. While these traits are defined by their extremes, in real life most cats are intermediate on one, two, or even all three. Insofar as their owners are concerned, there may be no such thing as an "average cat"—but many do in fact approximate to that description.

The bold/shy dimension is perhaps the most important of all, because it affects not only how the cat behaves on a minute-by-minute basis, but also how much and what each cat learns. In some situations, a bold cat learns more from a new experience than a cat that holds back. However, if the bold cat behaves overconfidently and is hurt as a consequence—for example, if it struts up to a belligerent tomcat—then not only may it get injured, but it may also learn less from its experience than a more circumspect cat that simply stands by and watches the encounter.

Whether a cat gets on with other cats, or is especially affectionate toward people, seems strongly influenced by its experiences as a kitten and during its adolescence. At least, we have yet to detect a strong and lasting genetic influence among crossbred cats. Scientists once thought that some cats carried "friendly" genes and others "unfriendly," but when they investigated this in more depth, they found that the differences were in fact due to genes that affected how bold the cats were. Bold and shy cats learn differently how to interact with cats and people. This is not to say that bold cats are necessarily friendlier than shy cats, or vice versa, although they are likely to express their affection in slightly different ways.

The notion that bold and shy cats learn in subtly different ways emerged from a classic experiment in which the offspring of two tomcats, one with a reputation for producing "friendly" kittens and the other "unfriendly," were raised in groups that were socialized in slightly different ways, some given minimal handling, while others were handled daily.[13] At one year old, the offspring were compared by placing them in an arena with an unfamiliar object: a cardboard box they'd never seen before. The offspring of the "friendly" father explored the box most rapidly and most thoroughly, and the offspring of the "unfriendly" father tended to hold back. The genetic difference between the two fathers thus influenced something more fundamental than just how friendly their kittens had become; it affected how their kittens reacted to anything they hadn't encountered before.

Similarly, how the kittens interacted with people during their socialization period was affected by their boldness. The kittens with the bold father approached people spontaneously, and as a result learned quickly how to interact with them. The shy kittens took longer to achieve the same level of confidence with people. However, given enough handling, these shy kittens could turn out just as friendly as the kittens with the bold father—though as expected, they tended to show their friendliness in a less "pushy" way than the kittens with the bold father. The shy father's kittens became fearful of people only if they were not handled every day: even at a year old these kittens, like their father, moved away from people, hissing and flattening their bodies to the ground. The amount of exposure to people they had received was, for the most timid, significantly less than the average kitten born in a typical home experiences, so the study did not precisely replicate normal conditions. However, it does provides a valuable insight into how vulnerable the offspring of shy tomcats may be to interruptions in their socialization.

Kittens born in someone's home likely receive enough handling to end up at least reasonably friendly to people, regardless of whether they are genetically bold, shy, or somewhere in between. The ways they show their affection may differ, however, and this seems to interact in a complex way with their early experience. In 2002, my team researched this interaction in a study on twenty-nine cats from nine litters born in regular homes.[14] The amount of handling the litters received in their second month varied from twenty minutes to more than two hours per day. When these kittens were eight weeks old, just before they were homed, we tried picking them up, one at a time. Those that had received the least handling were definitely the most inclined to jump down; we could hold those that had received the most handling for several minutes at a time. The amount of handling they had received seemed to have had more of an effect on their behavior than any genetic effects; all the litters had different mothers, and although we had no idea who their fathers were, the homes they'd been born in were far enough apart to make it unlikely that any two had the same father.

When we repeated our test two months later, when the great majority of the kittens had been moved to new homes, we found exactly the opposite: the kittens that had received the most handling during their second month of life were now the *most* restless, and those that had received the least were now the quietest.

This apparent contradiction probably demonstrates that not only does socialization to people in general start during the second month, so too does attachment to specific people. Very few of the cats appeared distressed when they were picked up at eight weeks old, so all must have received more than enough socialization to create a generally friendly kitten. Nevertheless, those handled the least were still not entirely confident when picked up by a stranger. Two months later, these kittens had gone through the process of learning about a new set of people—their new owners—and showed through their behavior that they were perfectly content to be picked up by anyone. The kittens that had received a great deal of handling in their original home may consequently have become extremely attached to their original owners, and therefore found the transfer to their new home particularly unsettling. Despite the two months of acclimatization that had passed, when we returned to test them again they may still have been anxious in these new homes.

The handling they'd received seemed to set these kittens' personalities off in different directions, but all effects of the differences in handling during kittenhood gradually disappeared as these cats matured. At one year of age, they varied in how much they liked being picked up by a person they hadn't seen for eight months, but this bore no relation to how much handling they'd received as kittens, or indeed to their genetics; cats that had originally been littermates were no longer similar to one another. When we tested them again at two and three years old, we found that they had changed little from the way they were at twelve months; so, by the end of their first year, each cat had already developed its own distinctive way of reacting to people.

Somehow, the way cats react to being picked up by unfamiliar people changes as they grow through their first year. This change is probably influenced by their new owner's lifestyle and how it interacts with their developing personality. However, once they are about one year old, their reactions vary much less. For example, some cats presumably become accustomed to being in a busy house and are unphased by strangers; others prefer the company of their owners, and hide when visitors stop by. The way cats arrive at these states of equanimity is affected by the amount of handling they receive before they leave their mothers, and almost certainly also by their genetics, but the end result appears to be roughly the same whatever the route they have traveled.

Most cats are extraordinarily sensitive to human body language, much more so than they usually receive credit for. This sensitivity enables them to adapt their behavior to the people they meet. People who dislike cats often complain that they are the first person in the room a cat makes a beeline for, so I decided to test this theory by staging encounters between cats and people who either liked cats or found them repulsive.[15] The people—all men, since we could not find any women who admitted to hating cats—were instructed to sit on a couch and not to move when a cat came into the room, even if it tried to sit on their laps. However, we could not prevent the cat-haters from looking away from the cat, which they usually did within ten seconds of first seeing it. The cats, for their part, seemed to sense the disposition of the people they were meeting within a few seconds of entering the room. They rarely approached the cat-phobics, preferring to sit near the door and look away from them. It was unclear how the cats were detecting the difference between the two types of men:

perhaps they could sense that the cat-phobics were tenser, or smelled different, or glanced nervously at the cats. Nevertheless, the cats' reactions show that they can be keenly perceptive when encountering someone for the first time. However, one of our eight cats, while apparently equally perceptive, behaved contrary to the other seven, singling out the cat-phobics for the most attention, jumping on their laps and purring loudly, much to their disgust. Cats such as this one presumably make a lasting impression on cat-phobics, rather than the majority of cats that sensibly avoid them.

Young cats especially seem to be much more adaptable animals than their detractors—and even some of their supporters—might have us believe. During the first year of their lives, they effectively tailor their personalities as much as they can to suit the particular household, or indeed other environment, in which they find themselves. Investigations on how they do this are still in their infancy, and the process is likely to be both drawn-out and contain features that are intrinsically private; still, intriguing links are emerging between cat behavior and their owners' personality. For example, owners who have intensely emotional relationships with their cats tend to have cats that are very happy to be picked up and cuddled.[16] This could be the cat simply adapting to the owner's demands, although studies have suggested that cats tend to resist being picked up when they're not ready for it. It's possible, therefore, that people who obtain an adult cat and want one that loves to be picked up deliberately select one with that kind of personality, rather than the cat changing its ways to suit its owner's demands. However, young cats and kittens are almost certainly more adaptable than older cats are.

A typical cat's relationship with its owner is apparently not rigidly determined by whether the cat is genetically bold or shy, even though both of these personality types persist in the general cat population. More important than its personality is the amount of handling the cat receives when it is a kitten, which alters the way it behaves for the first few months of its life; after this time, most cats successfully adapt their behavior to the demands of their new owners. Some kittens do not receive much handling, perhaps because they have been born in a cat shelter where the staff is overstretched, or because their mother, being shy herself, has hidden them away. If such kittens also have an inherited genetic tendency toward shyness, inherited from their mother or indeed their father, they may be at risk of never developing a fully affectionate relationship with their new owners. In all the studies that trace the development of kittens' personalities, some cats have disappeared from their owner's homes. While we have never been able to explain this fully, we have suspected that some were somewhat unsuited to being pets and had chosen to become feral.

Continued gentle interaction with people can buffer the effects of genetic shyness, so that the young cat learns to come out of its shell. If a cat that is genetically disposed toward shyness does not receive this handling, then the innate shyness may persist into adulthood. These cats, if allowed to breed, will still carry the "shy" genes they inherited from their parents; so if they mate with another "shy" individual, these genes will persist throughout the cat population.

We know little about how a cat's genetics affects its sociability toward other cats. The little existing research has focused instead on how a kitten's early experience of other cats alters its behavior as an adult. Kittens hand-raised on their own behave abnormally toward other cats, and kittens hand-raised alongside a littermate less so—nevertheless, all display a bizarre combination of fascination and fear on encountering another cat.[17] The presence of the mother, or in her absence another friendly adult cat, is apparently necessary for kittens to develop normal social behavior.

Cats raised in the normal way by their mothers also differ from one another in how friendly they are to other cats, albeit not exhibiting the extremes found in hand-raised cats. Much of this variation may also stem from different experiences during kittenhood: kittens born into extended families normally find it easier to learn social skills than those born to solitary mothers. For example, in parts of New Zealand, some feral cats live around farmyards, subsisting on a diet of rodents and scraps provided by the farmer, much like farm cats in the UK or the United States. However, because of less competition from native predators than elsewhere, other cats live in the bush nearby, feeding exclusively on prey they have hunted for themselves and adopting a way of life that must be much like that of undomesticated *Felis lybica.*[18] Males roam between the two populations, keeping them genetically mixed, but females seem to adopt their mother's lifestyle: those born on the farms stay there and share their mothers' territories, while those in the bush strike out on their own. Cats thus appear to inherit a form of social "culture" from their mothers that may have little to do with their genetics.

Even cats that have lived their whole lives in groups can differ markedly in how they react to other cats, hinting at genetic as well as cultural influences. In one study of two small indoor colonies (seven females each), researchers found that the individual cats varied consistently in their calmness while interacting with the others, and varied also in the extent to which they chose to be close to or keep away from them. These aspects of their personalities were distinct from how sociable these cats were toward people, or how generally active and inquisitive they were.[19] Each cat appeared to have worked out its own way of interacting with the cats around it, which was not simply a reflection of how "bold" it was.

Each cat's boldness would had affected the way it had initially approached other cats when it was first introduced to them, since "boldness" is a trait that underlies all first encounters with novel situations. However, in repeated encounters with the same individuals, each had developed a new and eventually stable personality trait, sociability to cats, which was unrelated to how bold they were. We can only guess at the reasons behind these emerging differences, since we know little of how such traits are formed. For example, if the outcome of just the first few encounters fixes each cat's strategy for dealing with cats permanently, then its personality might be profoundly affected by whether it happened to be smaller and weaker, or bigger and stronger, than the cats it met early on. However, if this aspect of personality develops over many encounters, then the subsequent differences between each cat could conceivably be driven by genetic factors, distinct from those that affect whether a cat is bold or shy.

Indeed, the domestic cat's capacity to live amicably with other cats should be genetically variable. Its wildcat ancestors seem incapable of extending their affiliations beyond temporary bonds between mother and offspring, but many domestic cats seem capable of forming affectionate bonds with adult cats. It seems highly unlikely that this change has already evolved as far as it can. Some cats, at least anecdotally, are unusually outgoing toward other cats, while others prefer to be virtually solitary. Dissimilarities in early life experience certainly caused some of these differences, but surely genetics also plays a part. We should still see variation between today's cats in how easy they find it to forge bonds with others. How shy or bold they are may affect this variation, although as with affection for people, boldness or shyness may have more influence on how the cat approaches another cat than on whether or not they become friends or enemies.

On the face of it, we would not expect much genetic variation underlying the domestic cat's ability to hunt. Not only is the cat descended from a specialist predator, but the primary function of the cat in human society has been, until recently, to kill rodent pests. Moreover, until the appearance of nutritionally balanced cat food some forty years ago, cats that were incompetent predators would not have bred as successfully as those that were. Scientific studies have confirmed that, given enough exposure to real prey, all cats have the potential to become competent predators by the time they are six months old. The way each cat hunts—whether it moves around constantly or sits and waits for hours where it knows that prey will appear—varies a great deal from individual to individual. So too does the type of prey in which each cat specializes: some cats seem incapable of catching a bird, and others catch more birds than rodents. We could regard these variations as aspects of "personality." However, these differences probably stem from each cat's experiences as it refines its hunting skills, since it will likely repeat tactics that have led to a meal. We have no evidence that some cats are born bird-hunters, and others innate mouse-catchers.

Surprisingly, research *has* revealed considerable differences among kittens in their prey-catching competence, particularly during their third month of life. By this stage in their development, all kittens are capable of performing their basic repertoire of predatory behavior—stalking, pouncing, biting, and raking with the claws—which they have practiced for the past several weeks through play with inanimate objects around the nest, as well as in mock-predation on their littermates. Despite all this practice, kittens between two and three months old vary greatly in effectively putting these actions together, in assessing what they can likely catch and what they should avoid, and in selecting the appropriate tactics for the prey in question—for example, not chasing after birds that are already flying away.[20] Three months later, however, they are all equally competent; somehow, the laggards catch up. Scientists have found no developmental reasons for the differences between kittens during their third month, so it is possible that these variations at least are genetically influenced. Genetics undoubtedly affects the rate at which kittens develop in general—for example, the age at which their eyes open—so this hypothesis is reasonable.

As every cat owner knows, cats differ not just on the outside. Inner and outer qualities are both affected by genes and the environment in which the cat grows up, but to different extents and in different ways. Cats' personalities develop according to a highly complex interplay between genetics and what the cat experiences during the first year or so of its life. These experiences can have extremely powerful effects that can all but obliterate any trace of genetics. Yet among cats that receive what is, for their species, a "normal" upbringing within a human household, signs of genetic effects on personality are apparent.

Cats also vary in terms of how quickly they learn to hunt. Although experience again plays a large part in determining this aspect of a cat's personality, it also seems likely that genetic factors are also at work. Cats are not only able to adapt as individuals to the circumstances in which they find themselves; they are also members of a species that contains a significant amount of genetic variation that affects their behavior, giving them the potential to evolve further as our demands upon them change. Today, the most significant challenge facing cats is a growing reputation as destroyers of wildlife, but even their most vociferous critics have to admit that not all cats are to blame. If hunting ability is linked to personality, and personality has a genetic basis, then it may become possible to predict which cats are likely to cause the least offense.

CHAPTER 10

Cats and Wildlife

Few topics get wildlife enthusiasts as riled as predation by domestic cats. In the wrong environment, cats can undoubtedly cause substantial damage to other species, especially where there is little competition in the shape of wild predators. Although many pet cats do undeniably go out hunting, it has proved remarkably difficult to pin down the impact of that hunting—whether it does have a significant effect on wildlife numbers. Indeed, when the balance of wildlife changes in a particular place, cats become convenient scapegoats. Well-fed pet cats should not need to hunt to supplement their diet, and in that sense any damage they do is unnecessary. Moreover, their habit of bringing their kill home rather than eating it on the spot does make it easy for their detractors to point the finger at them—it can seem as though they are killing for "sport"—while deaths due to wilder predators go unnoticed.

More insidiously, general anti-cat sentiments can creep into scientific literature addressing wildlife conservation. A group of scientists in Australia has recently called for "restrictions on the maximum number of cats allowed per household, mandatory sterilisation and registration of pet cats, curfews, requiring pet cats roaming outdoors to wear collar-mounted predation-deterrents or compulsory confinement of cats to their owners' premises," even though it is far from clear that any of these restrictions would lead to a recovery in the local wildlife.[1]

In 1997, the UK's Mammal Society produced an estimate of 275 million animals killed in Britain each year by pet cats. These figures derived from forms completed by their youth wing, "Mammalaction": data from the 696 cats surveyed were extrapolated up to the 9 million cats then in the UK. However, when the full analysis was finally published in 2003, it became clear that very few cats—less than 9 percent—had been included that did not hunt at all, even though most other studies had concluded that only about half of pet cats ever bring any prey home; fewer in urban areas, slightly more in the countryside.[2, 3] The reason for this bias appeared to be in the design of the questionnaire, which encouraged Mammalaction members to submit their results only if their cat had brought in some prey during the five months of the survey.

Despite these shortcomings, the figure of 275 million is still widely quoted by many influential organizations, including the Royal Society for the Protection of Birds, the British Trust for Ornithology, and the Bat Conservation Trust. When the figures were first announced in 2001, British wildlife TV presenter Chris Packham, a self-confessed "cat-hater," appeared on BBC radio describing cats as "sly, greedy, insidious murderers," and calling for them to be "shot"; more recently, he asserted that all cats should be given "ASBOs" (Anti-Social Behavior Orders, court-imposed restrictions on people likely to "cause harassment, alarm or distress to an individual householder or a neighborhood").[4] Similarly, when the University of Georgia's Kitty Cams revealed that a small minority of cats in Athens, Georgia, were killing a couple of lizards each week, the *Los Angeles Times*'s Paul Whitfield wrote, "So they're slaughtering wildlife, and you can't trust them. Seems like grounds for government action. . . . Present owners can keep their cats. But as the tabbies die off, so does cat ownership."[5] A January 2013 *New York Times* report of an estimate of predation by cats across the United States made by scientists from the Smithsonian Institute generated more email responses than any other story that day. The headline of another report on the same study read, "Domestic Cats Are Destroying the Planet."[6]

Looking at this situation more objectively, the impact cats actually have on wildlife varies enormously from one type of environment to another. The most dramatic effects undoubtedly occur on small, oceanic islands onto which cats have been introduced. Many of these islands contain unique fauna that have evolved due to their isolation from the mainland. Others are refuges for seabirds that raise their young in safety there, undisturbed by predators. When cats appear, they can cause havoc. Occasionally, these cats have been pets, perhaps the most notorious being the lighthouse-keeper's cat that killed the last specimen of the Stephens Island wren in 1894.[7] Usually, however, the cats have either escaped from visiting ships, or have been deliberately introduced to suppress pests such as rabbits, rats, and mice—usually accidental introductions in themselves. Lacking competition from other mammalian predators, cats can thrive in such environments, making easy prey of the local wildlife that has previously been unexposed to such hunters. With such an abundance of food available, cats breed prolifically, producing large populations of ferals. Perhaps counterintuitively, these feral cats sometimes do most harm on islands where they are not the only introduced animal: some researchers have suggested that an abundance of house mice, for example, provides the cats with a stable diet, thereby enabling them to increase in numbers to the point where they can exterminate the more vulnerable local wildlife.

We must keep the effects of feral cats in perspective, even when considering the island-cat situation. Island species account for 83 percent of all documented extinctions of mammals: isolated from many of the diseases, parasites, and predators that plague their relatives on the mainland, these species are intrinsically vulnerable. Yet scientists have been able to implicate feral cats in only about 15 percent of such extinctions, and even in these, other introduced predators have to take their share of the responsibility. Foxes, cane toads, mongooses, and especially rats can be equally, if not more devastating. Black ("ship") rats probably do more damage than any other introduced predator, and because cats are effective hunters of this species, their presence can sometimes even be beneficial.

For example, on Stewart Island off the coast of New Zealand, feral cats have existed for more than 200 years side by side with an endangered flightless parrot, the kakapo. These cats feed mainly on introduced species of rats (black and brown), which have been held responsible for the extinction of several other species of birds in the same region. Removal of the cats here might lead to an increase in the rat population, and could therefore potentially lead to the extinction of the kakapo.[8] However, we cannot deny that eradicating cats from islands has in some cases led to dramatic recoveries in the populations of threatened vertebrate species: examples include iguanas on Long Cay in the West Indies, deer mice on Coronados Island in the Gulf of California, and a rare bird, the saddleback, on Little Barrier Island in New Zealand.

On the mainland, feral cats can undoubtedly be effective predators in some locations, but their impact is much more difficult to quantify. Nevertheless, the sheer numbers of stray and feral cats in the world, somewhere between 25 and 80 million in the United States and around 12 million in Australia, suggests that they must have a major impact.[9] Feral domestic cats are "alien" predators throughout much of their range: indeed, they have been in the United States for less than

A feral cat stalking a kakapo

500 years. In most places where they have been introduced, cats seem able to compete quite effectively with local, "native" predators, even though the latter should be better adapted to local conditions.

Feral cats do have three advantages over other predators. First, their numbers are constantly added to by cats straying from the pet population, or emigrating from farms where cats are still kept for pest control. Second, because they are generally less fearful of humans than many wild carnivores, they can better take advantage of food accidentally provided by people, such as at garbage dumps, to sustain themselves when prey is hard to find. Third, because they resemble pet cats in everything but behavior, they attract the sympathy of many people, some of whom devote their lives to providing them with food and even veterinary care.

The most scientific attention to this problem, and perhaps the most public outcry, has occurred in Australia and New Zealand, where cats seem to be relatively recent introductions.[10] In both countries, many small marsupials and flightless birds have undoubtedly gone extinct, but the main culprit may be loss of habitat, not predation. Even where predators have been a major factor, the responsibility is often shared among cats, rats, introduced red foxes, and (in Australia) dingoes. According to Christopher R. Dickman of the Institute of Wildlife Research in Sydney, "the effects of cats on prey communities remain speculative."[11]

In some situations, cats can be a major cause of decline; in others, they may be protective. Predation from cats appears to have made a major contribution to the decline of some threatened native Australian species, such as the eastern barred bandicoot in Victoria and the rufous hare-wallaby in the Northern Territory. On the other hand, in a study of remnant patches of forest in suburban Sydney, the presence of cats *protected* tree-nesting birds, apparently because they themselves were hunting rats and other animals that would normally have raided the nests.[12] Cats can also suppress the numbers of introduced mammals, such as mice and rabbits, that compete for food with the native wildlife.

Despite the equivocal evidence, several Australian municipalities have pressed ahead with measures to reduce the impact of cats on wildlife. These include confinement of cats to owners' premises at all times, prohibiting cat ownership in new suburbs, nighttime curfews, and impounding free-roaming cats in declared conservation areas—even though only the last of these would control the activities of the feral cats that are probably causing the most damage.

Researchers have yet to evaluate the effectiveness of such control measures comprehensively. However, a recent survey of four areas of the City of Armadale, Western Australia, suggests that cats may not be the primary culprits after all. One area in this study was a no-cat zone, where cat ownership was strictly prohibited; the second was a curfew zone, in which pet cats had to be belled during the day and kept indoors at night; and in the other two areas, cats went unrestricted. The main prey species in the area were brushtail possums, Southern brown bandicoots, and the mardo, a small predatory marsupial a little bigger than a mouse, which was predicted to be the most vulnerable of the three to cat predation. In fact, researchers found more mardos in the unregulated areas than in the curfew or no-cat areas, and saw little difference in the numbers of the other two prey species across any of the sites. What variation there was could be best accounted for by the amount of vegetation available: in other words, habitat degradation, and not cats, may have been the major factor limiting the numbers of small marsupials. The draconian control measures against pet cats had, at least in this one location, produced no benefit to wildlife.[13]

How much long-term damage to wildlife do pet cats cause? Estimates of what proportion of pet cats ever kill anything vary considerably, but figures of between 30 percent and 60 percent seem reasonable, even when those cats kept indoors without access to prey are excluded. We have little reliable information on how many animals those cats that do hunt actually catch, because such events are rarely observed. What is usually recorded is not how many animals are caught, but how many are brought back dead to their owners—and then a "correction factor" is used to calculate the number killed, to account for those prey items that are eaten where they were killed, or simply discarded. The number brought home per cat is often quite low: for example, 4.4 animals per cat, per year in a recent UK study.[14]

The proportion of prey brought home has been calculated only twice, coming out at around 30 percent (although one of the two studies examined only eleven cats). Recently, a new study in the United States has provided a much more detailed picture, both metaphorically and literally, as the cats were fitted with Kitty Cams, lightweight video camcorders that provided a view of everywhere they went for a week or more. These cats took home around a quarter of their prey, they ate another quarter, and left the remaining half uneaten at the capture site. What may make this study somewhat atypical is that the main prey taken was a lizard, the Carolina anole, which many cats find unappetizing. In places where the main prey consists of mammals—such as the more palatable woodmouse commonly taken in the UK—both the proportion brought home and the proportion eaten might be higher.

Once we take these "correction factors" into account, and the figures scaled up to the whole of an area's cat population, the total number of prey taken can at first sight seem alarming. The Mammal Society's figures of 275 million per year might be overstating the case, but between 100 and 150 million may be a reasonable estimate for the entire UK. The recent Smithsonian study produced an estimate of between 430 million and 1.1 billion birds killed annually by pet cats in the mainland United States.[15] Furthermore, individual anecdotes, taken in isolation, do at first glance suggest that cats might be capable of bringing about local exterminations. For example, biologist Rebecca Hughes from the UK's Reading University reported that, in the cold winter of 2009–10, "One cat that lived beside a woodland brought in a blue tit [chickadee] every single day for a fortnight."[16]

Whether such levels of predation make any significant difference to wildlife populations in the long term is far less easy to assess. To take the blue tit as an example, the UK has an estimated 3.5 million breeding pairs, each producing seven or eight young per year—about

25 million more young birds than would be required to keep the population constant. So, some 25 million blue tits must die during the course of most years. Some do not make it out of the nest and some are victims of predation, but many starve during cold winters because their metabolism runs so quickly that they are barely able to store enough food to keep themselves alive overnight. Some of the birds brought home by the cat referred to above may well have died of natural causes during the night, and were then retrieved by the cat in the morning. In fact, the numbers of blue tits in UK gardens have increased by a quarter over the past fifty years, so pet cats are unlikely to be having any major effect on their numbers year on year. Most of the cat's favorite prey species are equally profligate breeders—in built-up environments in the UK, these include woodmice (fifteen to twenty young per pair per year), brown rats (fifteen to twenty-five), and robins (ten to fifteen).

Rather than contributing to the decline in wildlife populations, cats may simply be killing (or collecting) animals that are in any case not destined to survive for much longer—those that are ill or malnourished. Such animals are by their very nature the easiest to catch. One study that examined the condition of birds brought home by cats tended to confirm this idea: these birds were generally underweight and in poor condition.[17]

Recent evidence shows that garden birds may be evolving strategies to cope with cats. In rural areas, the main predators of European songbirds are usually the sparrowhawk and the kestrel; in urban gardens, the main predator is often the cat. By comparison with their country cousins, common birds living in urban gardens such as sparrows, robins, chickadees and finches were found to wriggle less, were more likely to "play dead," were less aggressive, and made fewer shrieks and alarm calls, all by comparison with their country cousins. The longer an area had been urbanized, the greater the difference, which suggests that the urban birds had not simply learned about avoiding cats, but had actually evolved a new set of defense mechanisms over the hundred or so generations since large-scale urbanization began in the mid-nineteenth century.[18]

What seems to irritate wildlife enthusiasts the most is not that pet cats hunt at all, but that they should not have to hunt, since the majority are well fed by their owners. Cats are therefore often portrayed as committing "murder," as opposed to other predators, which kill legitimately, to survive. Because they are fed, pet cats can exist at a much higher density than they ever could if they had to catch every meal for themselves. Thus, even occasional hunting could have a substantial impact, simply because there are so many cats.

Cats have not yet lost their desire to hunt; too few generations separate them from the rodent controllers of the nineteenth and early twentieth centuries. They will go out hunting even if well fed—in the past, when they were mousers, a single catch did not provide enough calories to allow them to relax between meals—but hunger does affect how intensely they hunt. Feral cats, even those that obtain most of their food from scraps, spend on average twice as much time hunting as pet cats do. Mother cats with kittens to feed will hunt almost continuously if they are not themselves being fed by someone.

Feral cats foraging in garbage

By contrast, pet cats rarely hunt "seriously," often watching potential prey without bothering to stalk it. A hungry cat will pounce several times until the prey either escapes or is caught; a well-fed pet will pounce halfheartedly and then give up, probably explaining why pet cats, when they do kill birds, usually succeed only when they target individuals already weakened by hunger or disease. Furthermore, pet cats rarely consume their prey, often bringing it home as if to consume it there, but then abandoning it.

The quality of a cat's diet also affects its desire to hunt. In one recent study conducted in Chile, cats fed on household scraps were four times more likely to kill and eat a mouse than cats fed on modern pet food. In another, cats eating low-quality cat food would break off to kill and chase a rat, but cats eating fresh salmon ignored the same opportunity to hunt.[19] These and other similar observations suggest that cats fed on scraps or nutritionally unbalanced cat food are strongly motivated to hunt, in particular by an impulse that they must supplement their diet to maintain their health. The domestic cat, in common with all its wild felid cousins, has highly specialized nutritional requirements that can be met from only one of two sources: either modern, nutritionally balanced commercial cat food, or prey (see box on page 71, "Cats Are the True Carnivores"). Scraps and low-quality cat food tend to be high in carbohydrates. If eaten day after day, carbohydrates seem to give cats a craving for protein-rich food, which in their world means flesh. Commercial pet foods are of much better quality than they were half a century ago, so most modern pet cats have likely received a nutritionally balanced diet every day since they were weaned, and so are unlikely to become prolific hunters. Cats that have been neglected or strayed at some time in their lives may have been driven to hunt through nutritional necessity: once they have acquired the habit, it may be difficult to lose, so such cats may require extra precautions to prevent them from hunting unnecessarily (see box on page 251, "How Can We Prevent Cats from Hunting?")

Of course, we all want to minimize the damage cats do to wildlife. The best approach to this problem will vary depending on what kind

How Can We Prevent Cats from Hunting?

Surveys show that the large majority of pet cats catch very few birds or mammals. If your cat is part of this majority, then you need not take any countermeasures, unless you happen to live next door to a nature reserve.

If your cat is a prolific hunter, one of the following may reduce its impact:

- Equip your cat with a belled collar. Although some studies have found little effect, several have shown significant reductions in the number of prey caught, both mammals and birds.
- Add a neoprene bib to your cat's collar. This interferes with its ability to pounce and may reduce the number of birds that it catches.[20]
- Add an ultrasonic device to your cat's collar to warn potential prey of its approach.
- Keep your cat indoors at night. This may reduce the number of mammals it catches, and slightly reduces its own risk of being hit by a car.
- Play with your cat, and allow it to "hunt" prey-like toys. This may reduce its motivation to hunt, although this has never been scientifically evaluated.

Any collar you put on your cat should be of the snap-open type, as other types could strangle the cat; see www.fabcats.org/owners/safety/collars/info.html for additional information.

Additionally, cat owners can mitigate any damage their cat might cause to the local wildlife by taking positive steps to provide food and refuges. Feeding birds at a cat-proof bird table and building a log pile in the corner of the garden as shelter for small mammals are just two measures that counteract any effect of the cat's hunting.[21]

of cat we are dealing with, and particularly how closely the cat in question is associated with people. Pets and ferals require different solutions. On oceanic islands, cats are invariably introduced "aliens," largely or completely unsocialized to people, and occupying a niche that would otherwise have remained vacant, since medium-sized land mammals cannot reach such places without man's help. Whether the cats were introduced deliberately, or are the descendants of escaped ships' cats or settlers' pets does not matter; either way, they are now essentially wild animals. Eradication of these cats, by humane means, is usually the only means whereby each island's unique ecosystem can be restored, although this should not be carried out in isolation: other "alien" species, such as rats and mice, may themselves decimate the local fauna once they are liberated from the pressure of being preyed on by cats. Perhaps surprisingly, given the publicity that the damage that cats can do receives, only about 100 such eradications have taken place so far, with thousands more islands still affected by the presence of feral cats.[22] Ultimately, however, widespread humane eradication may be the only way to fully restore fragile island ecosystems.

Minimizing the impact of feral cats on wildlife is much more difficult when they live near pet cats—which, on the mainland, is almost everywhere. We have few reliable estimates on how much damage such feral cats do, mainly because they are rarely the only predator present, competing with both their native equivalents and also introduced species such as the red fox and the rat. Feral cats, even those that obtain some of their food from handouts or scavenging, are of necessity much more "serious" hunters than the vast majority of pet cats, and so per capita must be responsible for more damage to wildlife.

In many locations, human activity has reduced areas of conservation interest ("biodiversity value") into small "islands," albeit islands surrounded by concrete rather than water. For example, urbanization has broken the once-contiguous heathland habitat of the sand lizard on the south coast of England into fragments, making each isolated population highly vulnerable to extinction by wildfires. In other, similarly fragmented habitats feral cats could potentially cause considerable damage, although well-documented examples remain scarce. Eradication of feral cats from areas where they coexist with pet cats is both problematic and ultimately unproductive. Unless all pet cats are curfewed, or kept permanently indoors, or compulsorily registered and microchipped, it is virtually impossible to be certain that a cat that has been trapped is feral, particularly if it is somewhat socialized to people. Even if local eradication was achieved, the niche formerly occupied by the feral cats would still exist, and would soon be filled by stray cats or by ferals migrating from other areas.

Although they rarely say so outright, it is difficult to avoid the impression that conservationists and wildlife enthusiasts would like to exterminate all feral cats. This would account for their vehement objection to Trap-Neuter-Return (TNR) schemes, in which feral cats are, for welfare reasons, neutered and then returned to the site where they were originally trapped. Although such schemes might, in theory, eventually lead to the disappearance of the feral cat population in the locality in question due to reduced breeding, this has rarely actually been achieved. Undoctored cats migrate into the area to which the neutered cats have been returned, and soon replace the original breeding capacity. Indeed, such schemes can inadvertently generate "hotspots" for the abandonment of unwanted cats, their owners believing, possibly mistakenly, that the cat will join the feral colony and be allowed to share its food. Even where the site is fairly well isolated and trapping and neutering is maintained for years, the feral cats rarely disappear completely.

I studied one such colony that had formed around a partly derelict hospital in the south of England, originally built in the nineteenth century as an insane asylum, and—as such places usually were—located several miles from the nearest village. Prior to the introduction of TNR, the colony had consisted of several hundred cats and kittens; several years later, numbers had reduced to about eighty cats. Many of these were members of the original colony, now neutered and gradually becoming elderly, but at least one tomcat and several females had evaded capture and were continuing to breed. In addition, pregnant females would appear periodically, presumably having been "dumped" there—although, being well socialized, these were easy enough to trap and re-home via a humane charity. Overall, the colony stayed the same size, residual breeding and immigration replacing those of the original group that were dying of old age, sustained by the food provided by the dwindling band of remaining long-stay patients.

Supporters of TNR maintain that once a colony has been neutered and its food supply stabilized, its impact on wildlife should reduce.

Neutered feral cats at the hospital

However, little reliable evidence has been found to support this position. The cats presumably continue to hunt: better nutrition may result in less consumption of prey items, but they, having caught the habit of hunting every day, nevertheless continue to kill and harass wildlife. However, the neutered cats do at least occupy space that, had they been euthanized rather than neutered and returned, would have soon become occupied by other cats. From the perspective of assigning finite resources to wildlife conservation, it may be better for conservationists to allow cat enthusiasts to assist in managing a cat population, if it is not causing catastrophic damage to wildlife. The much-touted alternative, the extermination of feral cats on a regular basis, is likely to alienate those of their supporters who care equally for cats and wildlife.

Since feral cats present such a slippery target where they coexist with free-roaming pets, conservationists tend to concentrate their efforts on restricting owned cats—this, despite an almost complete lack of hard evidence that pet cats are causing significant and lasting damage to wildlife. In an apparent attempt to fill the loophole caused by this lack of evidence, scientists at the UK's University of Sheffield have proposed a "fear effect" hypothesis: that pet cats suppress breeding in bird populations, their very presence triggering fear responses in the birds that inhibit foraging and depress fertility.[23] However, this theory discounts that urban birds seem to have evolved strategies to overcome the impact of cats. Furthermore, the mere presence of one lazy and ineffective predator must surely have less impact than the fear engendered by rats, magpies, crows, and other "serious" predators of small birds and their young.

When targeting pet cats for criticism, bird enthusiasts often fail to mention that other predators may make much more of an impact on bird numbers than cats do. Magpies, major predators of songbird nests, have tripled in number in the UK since 1970 to between 1 and 2 million individuals—with most of the increase occurring in towns where, coincidentally, cats have also increased. The Royal Society for the Protection of Birds—pledged to protect magpies, as well as their prey—investigated this increase in case it was tied to reductions in songbird populations over the same period. They concluded:

> The study . . . found no evidence that increased numbers of magpies have caused declines in songbirds and confirms that populations of prey species are not determined by the numbers of their predators. [Presumably, these include domestic cats, although they don't say so specifically.] Availability of food and suitable nesting sites are probably the main factors limiting songbird populations. . . . We discovered that the loss of food and habitats caused by intensive farming have played a major role in songbird declines.[24]

Magpie killing a blackbird chick

Cats rarely catch magpies, but they may inadvertently assist smaller birds by suppressing the populations of some of their other enemies. The UK has at least ten brown rats for every cat, and while rats are omnivores, their impact on bird and small mammal populations around the world has been well documented. Moreover, since young brown rats are among cats' favorite prey items, if cats suppress rat populations in towns, they may indirectly be helping birds.[25] Cat owners may therefore be able to do more for wildlife by improving habitats for small birds (and mammals other than rats) in their gardens than by confining their cats indoors (see box on page 251, "How Can We Prevent Cats from Hunting?").

Such precautions may not be enough to silence the cat's most vocal critics. Furthermore, most of today's owners revile or endure, rather than admire, their cat's hunting prowess. Unlike their ancestors, pet cats no longer need to hunt to stay healthy, so a reduction in their desire to hunt will do them no harm. Ideally, the cat of the future will have less inclination to hunt than the cat of today.

CHAPTER 11

Cats of the Future

We have more pet cats today than at any time in history. Over the past half century, the growth of organizations devoted to re-homing unwanted cats, together with advances in veterinary medicine and in nutritional science, have ensured that today's cats are far healthier than cats as a whole have ever been. Despite these auspicious trends, we now also see signs that their very popularity is adversely affecting their well-being. These effects will increase in the coming decades, so we cannot take the future of the cat for granted.

Humans are tending to cats' physical needs to an extent never experienced before. However, cats' emotional needs are still the cause of widespread misapprehensions. Cats are widely perceived as being far more socially adaptable than they actually are. Owners polled for a recent survey said that half of pet cats avoid (human) visitors to the house; almost all pet cats either get into fights with cats from neighboring houses, or avoid any contact with them; and half of the cats that share households with other cats either fight or avoid one another.[1] Research confirms that cats find such conflicts highly stressful: they experience fear during the event itself, and anxiety in anticipation of the next encounter. They are constantly hypervigilant through cues we are unaware of, such as the odor of a rival cat. Chronic anxiety can lead to deteriorating health and may reduce life expectancy. Unfortunately, we do not know enough about how to mitigate this situation, made worse by the ever-increasing number of cats kept as pets.

Owners also face increasing pressure to keep cats indoors, either permanently or just at night. Charities concerned with cats' welfare point out that urban environments present many hazards, including injury from traffic, wounds from fighting with other cats, exposure to diseases and parasites, and accidental poisoning. Most vocal are conservationists and wildlife enthusiasts who call stridently for cats to be kept indoors to prevent them from killing wildlife. Perhaps surprisingly, we have very little research on whether cats are adversely affected by being kept in one small area for all or part of the day, although it seems that some cats adapt much more readily to confinement than others.

Looking ahead, there has been virtually no discussion anywhere on what the cat of the future might be like. We seem to share an unvoiced assumption that because cats have always been around, they always will be; but as discussed in previous chapters, their circumstances are changing rapidly, and we cannot take their continued popularity for granted.

A century ago, the world was often, by today's standards, excessively cruel to cats, even those lucky enough to be chosen as pets; tying a lighted firecracker to a kitten's tail was considered amusing, and "kicking the cat" was so unremarkable as to pass into the vernacular as a metaphor for releasing frustration. Sentiments have changed radically since then: for example, in 2012 two Las Vegas teenagers accused of drowning two kittens in a cup of water were charged with animal cruelty, now considered a felony in Nevada.[2] We now have the resources to minimize cruelty—even if this is not uniformly exercised—and at the same time to curb cat populations by humane means.

Cats are prolific breeders. Left to their own devices, females produce many more kittens than are needed to keep the population stable, and in the past most of these kittens would have died before reaching adulthood, their lives not only short but also fairly miserable. Indeed, many unwanted kittens were drowned by their owners, and many of those that survived succumbed to debilitating respiratory diseases for which no vaccine yet existed. Over the past few decades, humane charities have promoted the use of neutering as the method of choice for restricting the numbers of house cats, setting themselves the goal of ensuring that every cat and kitten should find a loving home. To date, only a few more affluent locations have achieved this goal. This can be compensated by the transfer of cats from areas where neutering rates are low, so that they can be re-homed in areas where cats—or at least young, appealing cats—are in short supply. Some owners, now increasingly branded as irresponsible, still allow their young female cats to have one litter before they get them spayed. Others are caught unaware because on modern diets, females can conceive at six months old, whereas many owners do not get around to thinking about neutering until towards the end of their cat's first year. Other kittens may enter the pet population when stray females are rescued when they are pregnant or are found with newborn litters. Conversely, undoctored tomcats are rarely kept as pets nowadays, at least in urban areas (raising the question as to where the six-month-old females find fathers for their litters). For the time being, we have more cats available for adoption than owners wanting them, but as neutering becomes ever more widely practiced, kittens, especially, can become difficult to find.

If at some time in the future random-bred cats do become hard to find, then prospective owners will presumably turn to pedigree cat breeders, who currently supply no more than 15 percent of pet cats, and less than 10 percent in many countries. Luckily for cats, few of the mutations that have produced such extremes of appearance in the domestic dog seem to have been incorporated into cat breeds. The few that have, such as the dwarfing gene responsible for the munchkin cat's short legs, have been carefully scrutinized by geneticists anxious that cats should not go down the same road as dogs. However, some of the most popular pedigree cat breeds have begun to succumb to the afflictions of breeding for appearance, as well as the side effects of too much inbreeding (see box on page 260, "Pedigree Cats: The Dangers of Breeding for Extremes"). Moreover, the behavior of pedigree cats, while it differs somewhat from that of random-bred house cats, provides little variation that is not already found in random-bred cats, a situation quite different from that in dogs, where many breeds were originally derived for their behavior, not their appearance. Simply

Pedigree Cats: The Dangers of Breeding for Extremes

In the past few years, the media has paid much attention to problems created for pedigree dogs by indiscriminate breeding for looks.[3] The same plight for cats has been less newsworthy, but similar issues might appear in the future—indeed, some are already evident.

Breeding for appearance can create two problems. First, breeding for exaggerated features may cause the animal distress or result in chronic ill health. Among cat breeds, the classic example is the snub-nosed or peke-faced (technically, "brachycephalic") Persian. Traditionally, Persian cats have faces that are somewhat rounder ("doll-faces") than those of ordinary cats, and were derived from a variety of long-haired breeds including the Turkish Angora. The mutation that causes the flat face appeared in the 1940s in the United States, and was quickly adopted as the ideal for the breed. Further selection led to the nose becoming even shorter and higher in the skull, to the point where the nose became squashed between the eyes; this extreme form is now discouraged by the breed clubs. All brachycephalic cats are prone to breathing difficulties, eye problems and malformed tear ducts, and difficulties when giving birth, with a high proportion of stillborn kittens. Pet owners today seem to prefer the traditional style of Persian, as reflected in a fourfold drop in registrations of peke-faced Persians in the UK between 1988 and 2008.

A peke-faced Persian

Other breeds face health concerns as a result of breeding: surprisingly, one, the Manx, has been exhibited in cat shows for more than a century. The gene that gives the Manx cat its stumpy tail is essentially a defect,

(continues)

Pedigree Cats: The Dangers of Breeding for Extremes *(continued)*

often lethal: a kitten that has inherited two copies of this gene, one from its father and one from its mother, will likely die before birth. Cats with one copy of the gene grow tails of varying lengths, and some with partial tails are prone to arthritis that may produce severe pain. The gene can also affect the growth of not just the tail but also the back, damaging the spinal cord and causing a form of spina bifida. Manx cats are also prone to bowel disease.

A different skeletal malformation characterizes the "squitten" or "twisty cat" (not a recognized breed), which has incomplete development of the long bone in the front legs, resulting in the paw being twisted and attached at the shoulder, a deformity that has been likened to the effects of Thalidomide on human infants. Such cats cannot walk, run or dig properly, and cannot defend themselves; they do, however, sit upright in a "cute" manner, which presumably accounts for their "twisted" appeal.

In other breeds, problems caused by breeding for appearance are less obvious. For example, the gene that gives the Scottish Fold its characteristic lop-eared appearance also causes malformations of the cartilage elsewhere in the body, and as a result many of these cats develop severely painful joints at a relatively early age.

The second type of problem arises as a side effect of so-called line breeding, which is effectively inbreeding. The quest for the perfect specimen can result in the perpetuation of genes that disadvantage the cats that carry them—genes that would be quickly selected out if they appeared in alley cats, because they must impede hunting. For example, many Siamese cats have poor stereoscopic vision due to a lack of nerves in their brains to compare signals from the left and right eyes. As a result, they may see double, or one eye may shut down completely, sometimes causing a squint to develop. Another separate malformation in the retina leads to their vision blurring every time they move their heads. In wishing to sustain the Siamese cat's distinctive appearance, breeders inadvertently allowed this defect to continue from generation to generation.

replacing random-bred cats with cats from today's pedigree breeds will not only perpetuate those genetic problems that already exist, it also cannot solve the problems that the cat is facing as a species.

A reduced motivation to hunt and kill prey is just one of several factors that will enable cats to adapt better to twenty-first-century living. Allowing a little anthropomorphism: if cats could write themselves a wish list for self-improvement, a set of goals to allow them to adapt to the demands we place on them, it might look something like this:

- To get along better with other cats, so that social encounters are no longer a source of anxiety.
- To understand human behavior better, so that encounters with unfamiliar people no longer feel like a threat.
- To overcome the compulsion to hunt even on a full stomach.

The corresponding requests from owners:

- I'd like to have more than one cat at a time, and for my cats to be company—not just for me, but also for one another.
- I wish my cat didn't disappear into the bedroom to urinate on the carpet every time I have visitors.
- I wish my cat didn't bring gory "presents" through the cat-flap.

We currently know of two ways to achieve these goals. First, we can train individual cats to change how they interpret and react to their surroundings. The advantage of this approach is that its effects would be immediate; the disadvantage is that it must be repeated for each successive generation of cats. Second, because the cat's genome is not yet fully domesticated, there is still scope to genetically adapt their behavior and personalities to twenty-first-century lifestyles. The benefits of selective breeding toward our goals will become apparent only after several decades, but these changes would be permanent.

Cats are intelligent, and (up to a point) adaptable animals, so we can achieve some of these goals through directing cats' learning—

providing them with the right experiences to enable them to conform to the demands placed upon them. This will almost certainly involve a certain amount of formal "training." Although most dog owners know that they must train their dogs to make them socially acceptable, such a thought scarcely crosses cat owners' minds, or if it does, they reject it as only appropriate for "performing cats." Providing the right sort of experiences during the first few months of a kitten's life likely produces long-lasting effects, considering that this is the time when cats' personalities are forged, but more research is needed into precisely what those experiences should be.

The genetic basis for the cat's behavior must have changed as it adapted to living alongside man, even though the details are now lost in prehistory and ultimately impossible to trace. The average cat's personality is likely still changing, as some personality types fit modern conditions better than others. However, such haphazard change will not come about quickly enough to keep up with the pace of change that we now demand of our pets, as our own lifestyles change, so more direct intervention will be needed if the cat is to adapt at an acceptable pace.

We value our cats for their affectionate behavior, yet this trait has rarely been deliberately bred for—and then, only as an afterthought. This trait must have been accidentally selected for in the past, as loving cats get the best food![4] Nevertheless, "unfriendly" tendencies have persisted even among pet cats: in the 1980s experiments that defined the cat's socialization period, the experimenters noted "a small but constant percentage of the cats (about 15 percent) seem to have a temperament that is resistant to socialization."[5] Even today, cats seem to possess a large range of variation in the genes that underpin temperament, learning, and behavior, providing the raw material for work on selective breeding for behavior and "personality," rather than just appearance.

As cat owners, we have various resources available to help individual cats adapt to today's crowded conditions, but many people seem unaware of many of these. Our challenge is to use these tools early in the cat's life, when it is still learning about its surroundings. The second and third months of life are the crucial period, during which the growing kitten learns how to interact socially—with other cats, with people, and with other household animals. As we have seen, it is during this period of its life that the cat learns both how to identify its social partners, and how to behave toward them in a way that will produce the desired outcome—a friendly tail-up, a lick behind the ear, a bowl of food, or a cuddle. More generally, this is also the time when cats learn how to cope with the unexpected—whether to be curious, accepting the risk of approaching and inspecting new things, or whether to play it safe and run away. Research shows that a cat's capacity to take risks can be subject to a strong genetic influence, but learning must also play a part.

A cat that has had only limited exposure to different kinds of people during its second and third months of life may become timid, retreating to its safe place whenever people it regards as unfamiliar come to the house. Without exposure to people before eight weeks of age a cat will become fearful of humans in general. However this is not the end of the process of socialization, just the necessary beginning: kittens must have the opportunity to make their own connections with different types of humans.

Many kittens homed at eight weeks may miss out on useful socialization experience with other kittens. The third month of life is when play with other kittens peaks, and feral kittens maintain strong social links with their peer group until they are about six months old. Veterinarians often (sensibly) advise that kittens should be kept indoors for the first few weeks after homing, to prevent them from straying: however, if there are no other cats in the house, they may miss out on a crucial phase in the development of their social skills.

Individual cats adopt different strategies when encountering the unfamiliar. Many withdraw, hide, or climb to a safe vantage point. A minority may become aggressive, perhaps those that have not been given the opportunity to retreat on previous occasions. Their owner may have run after them and scooped them up rather than allow them to withdraw. These cats quickly learn that the stress of an undesired encounter can be prevented by scratching, hissing, and biting. Some of these cats develop this tactic even further, preemptively striking out at people they don't know or who have previously been forced on them.

Cats also develop their own preferred tactics for dealing with other cats they do not know. When making their first social encounters as kittens, some simply flee; others attempt to stand their ground, and often get a swipe or worse for their efforts. Few attempt to engage in a friendly greeting, and even fewer find such a greeting reciprocated. Flight or fight thus often becomes the young cat's default response when meeting unfamiliar cats.

Owners who wish to add another cat to their household have an opportunity to manage the introduction to have a positive outcome for all concerned. We cannot take for granted that cats will immediately like one another. The new cat will likely feel stressed at being suddenly uprooted from its familiar surroundings and dropped into what it perceives as another cat's territory, and the resident cat will probably resent the intrusion. Therefore, it is usually best to start by keeping the new cat in a part of the house that the resident cat rarely uses, allowing it to establish a small "territory" of its own and getting to know its new owners before facing the challenge of meeting the resident cat face to face. The two cats will undoubtedly be aware of each other's presence, if only by smell, but this will be less stressful at this early stage than being able to see each other. Owners can build up some degree of familiarity between the two cats before a meeting takes place by periodically taking toys and bedding from each of the two cats and introducing them to the other while rewarding that cat with food treats or a game. This builds up a positive emotional link with the other cat's odor. The actual introduction should wait until after both cats no longer show any adverse reaction to the other's smell, and should be carried out in stages, starting with allowing the cats to be together for just a few minutes.[6]

Because cats carry a reputation for being untrainable, most owners are unaware that training can reduce the stress that cats can feel in situations where they would much rather run away. For example, owners can use clicker training (see chapter 6) to entice a cat to walk into its cat carrier, rather than forcing it in.[7] Similar training could help cats overcome their initial fear of other potentially stressful situations—for example, encounters with new people. In general, cats need persuasion, not force, if they are to adopt a calm approach to new situations. If more owners understood the value of training, a great deal of stress could be avoided—for the cats, certainly, but also for the owner, if the cat's stress results in deposits of urine or feces around the house.

Training can also help cats to adapt to indoor living. Training a cat is a one-on-one activity that is mentally stimulating for the cat, and that also enhances the bond between cat and owner. It may also be useful in reducing some of the negatives of keeping a cat indoors. Cats need to express their natural behavior, and many owners understandably object to the damage to household furnishings that their cats unwittingly cause, for instance when sharpening their claws. In some countries, veterinarians remove the cat's claws surgically, but this may not be in the cat's best interest, and in some countries this intervention is illegal (see box on page 267, "Declawing"). A cat without claws may not only experience phantom pains from its missing toes, but is also unable to defend itself if attacked by other cats. Training the cat to claw only in specific places is a far more humane and straightforward alternative, especially if the cat has not yet developed a preference for soft furnishings.[8]

Although untested so far, training might also be useful in reducing cats' desire to hunt—or at least curtail their effectiveness as predators of wildlife. We know that when cats appear to be playing with toys, they think they're hunting, but we have no information on whether playing in this way reduces—or, just conceivably, enhances—their desire to hunt for real. If play does have an effect on this desire, how long does the effect last? Would a daily "hunting" game between owner and cat save the lives of garden birds and mammals? Is it possible to train a cat to inhibit its pounce?

We also know little about how experience affects the hunting habit in general. Cats vary enormously in how keen they are to go out

Declawing

Cats instinctively scratch objects with their front claws. Perhaps they do this to leave behind an odor or a visible sign of their presence, to alert other cats. They may also scratch because their claws are itchy: periodically, the whole of the outside of the claw detaches, revealing a new, sharp claw within. If this is not shed, maybe because the cat is arthritic and finds scratching painful, the whole claw may overgrow and cause the pad to become infected.

Some owners who object to scratch marks on their furniture seek to have the cat's front claws (and occasionally, too, its back claws) removed. Few veterinary procedures excite as much controversy as declawing (known technically as onychectomy). This is regarded as routine in the United States and the Far East, but is illegal in many places, including the European Union, Brazil, and Australia.

Declawing is a surgical procedure that involves amputation of the first joint of the cat's toes. The initial pain resulting from the procedure may be controlled with analgesics, but we do not know whether cats subsequently feel phantom pain due to the nerves that have been severed. However, cats and humans have nearly identical mechanisms for feeling pain, and four out of five people who have fingers amputated have phantom pain, so cats likely do as well. (I myself experienced phantom pain for more than ten years after most of the nerves in one fingertip were severed in an accident. I learned to ignore the pain because I knew it was meaningless—something cats are unlikely to be able to do.) Declawed cats are more likely to urinate outside their litter boxes than other indoor cats, possibly because of the stress of this phantom pain.

Claws are an essential defense mechanism for cats. While owners of indoor cats will argue that their cat never meets other cats, and so should never need their claws, a declawed cat that is picked up roughly by a person may resort to biting, unable to scratch to indicate its discomfort, and thereby cause a much more significant wound.[9]

hunting. That the basis for this is mainly genetic is unlikely, since only a few generations have elapsed since all cats had to hunt to obtain the right kind of food. Anecdotally, one of the arguments for allowing a female house cat to have one litter is that this is usually born in the spring, distracting the cat (provided its owners feed it well) from going out hunting and thereby learning its trade. Is there a "sensitive period" for perfecting predatory skills, after which the desire to hunt is unlikely to develop fully? Further study of this might not only save animals' lives, but also a great deal of aggravation between cat and wildlife enthusiasts.

Today's cats find themselves in a delicate situation. On one hand, they must adapt to meet our changing needs; on the other, they have a reputation for being a pet that is easy to maintain. Persuading many cat owners to train their cats, to spend that extra time and effort to change their pets' behavior, may be difficult. As such, we must focus on a genetic shift as well, taking the cat further down the road toward full domestication.

Ideally, cats that are predisposed to adapt to modern living conditions—to achieve the three goals outlined above—should be identified and then prioritized for breeding. This will not be entirely straightforward, since cats' personalities continue to develop after the normal age of neutering, so before any such breeding program can begin, research will be needed to separate the effects of the desired genes from those of the cat's social environment. Moreover, there is probably no single "perfect" set of genes that will enable cats to fit all the lifestyles that humankind will demand of them. The ideal indoor cat will almost certainly be genetically distinct from the ideal outdoor cat, since, among other differences, the owner can have far more influence on relationships with other cats if the cat is confined indoors.

We have three potential sources for such genes: house cats, pedigree cats, and hybrids. Conventional pedigree cats have been produced almost exclusively for their looks, not their behavior, so they are unlikely to be a rich source of new behavior traits.[10] Pedigree cats are derived entirely from cats that have only ever had two functions: to hunt vermin and to be good companions. In most breeds, there seems to have been little direct selection for behavior, only looks. There are however a few interesting exceptions.

The Ragdoll is a semi-longhaired breed that was originally named for its extremely placid temperament. The first examples to be exhibited, bred in the early 1990s, went limp when picked up, almost as if the scruffing reflex was triggered by touch anywhere on the body, not just the back of the neck. It was once rumored that these cats were insensitive to pain, and animal welfare organizations raised concerns that people might be tempted to toss these cats around like cushions. The breed no longer shows such extreme behavior, but is still renowned for its easygoing temperament. A similar breed derived from the same original stock, the RagaMuffin, is described thus: "The only extreme allowed in this breed is its friendly, sociable and intelligent nature. These cats love people and are extremely affectionate."[11] The genetic basis for these cats' relaxed sociality, at least toward people, is unknown, but could potentially be transferred to other cats by crossbreeding. Unfortunately for their welfare, ragdoll-type cats are reputedly vulnerable to attack from neighborhood cats, perhaps because they are simply too trusting, and for this reason many breeders advise prospective owners to keep them indoors.

Hybrids, crosses between *Felis catus* and other species, while initially produced mainly for their "exotic" appearance, have brought new genetic material into the domain of the domestic cat. Their behavior is often quite distinctive, so hybrids may potentially provide a source of genes that influence behavior that are not currently found in ordinary domestic cats.

The most widespread of these hybrids, the Bengal cat, may not offer any solution to the adapting the domestic cat to the twenty-first century, since its personality appears to have headed back toward that of its wild ancestors. The Bengal is a hybrid between domestic cats and the Asian leopard cat *Prionailurus bengalensis*. The latter species is separated by more than 6 million years of evolution from the domestic cat, and has never been domesticated in its own right; therefore, it would seem an unlikely starting point for a new breed of cat, were it not for its attractive spotted coat (referred to as "rosettes"). Domestic cats and Asian leopard cats will mate with one another if given no other option, but the resulting offspring are essentially untamable. During the 1970s, repeated breeding between these hybrids and domestic cats produced some offspring that retained the leopard cat's spots on the back and flanks, creating the current Bengal "breed."

Unfortunately, many Bengals possess not only wild-type coats, but also wild-type behavior, as this information from the Bengal Cat Rescue website confirms:

> This breed has a strong and sometimes dominant personality and although affectionate, lots are not simple lap cats. They can respond aggressively to discipline and to being handled . . . Their commonest problems are aggression and spraying. . . . Also hardly a week goes by when someone doesn't contact me about having bought or adopted a pair that are trying to kill each other. . . . Bengals enjoy climbing and this includes your clothes and curtains. They like exploring and are no respecter of ornaments or photographs. Often cat-aggressive, many will terrorise not just their own household but can actively seek out neighbors' cats and enter their homes to hurt them. They are not playing, they mean it.[12]

From the perspective of producing a docile pet, the Asian leopard cat was never going to be a good candidate for hybridization. This species is one of the few small wild cats that is not threatened with extinction; nevertheless, many zoos keep one or two specimens. These animals are, however, virtually untamable: zookeepers report that they are impossible to approach, let alone touch.[13]

From the perspective of seeking genes that could be useful for changing the domestic cat, some of the smaller South American cats could be better candidates for hybridization. In particular, Geoffroy's cat, similar in size to the domestic cat, and the slightly larger margay, are often friendly to their keepers when kept in zoos, and might therefore provide genetic material useful to the continued evolution of the domestic cat. The South American cats lost one pair of chromosomes soon after they diverged from the rest of the cat family some eight million years ago; this *should* mean that the offspring of a domestic cat and a South American cat is sterile, but surprisingly, hybrids with Geoffroy's cat can be fertile. The resulting "breed," usually known as the "Safari," was first created in the 1970s, but is still rare: it is usually still produced by mating the two original species, and the kittens, which grow into extra-large cats, fetch thousands of dollars. Breeding between these first crosses and ordinary domestic cats, the method

Bengal cat (above) and Safari cat (beneath)

used to produce the Bengal, seems to produce fertile offspring, but breeders apparently chose not to pursue this option. Hybrids with the margay, once referred to as the "Bristol," suffer from fertility problems and are apparently no longer bred. The margay is a tree-dwelling cat with double-jointed ankles, enabling it to climb down tree trunks as easily as other cats climb upward, to hang one-footed from branches like a monkey, and to leap twelve feet from one perch to another: such agility, if passed onto its hybrids, might be appealing, but also excessive for most domestic purposes.

Several other types of cat have been produced through hybridization with other felids, but all have been bred for their "wild" looks, and none seem to offer more than curiosity value to the domestic cat's genome. These include the Chausie, a domestic cat crossed with the jungle cat *Felis chaus*, and the Savannah (serval *Leptailurus serval*), as well as many other oddities of doubtful provenance. Some are classified as wild animals rather than as pets, much as wolf-dog hybrids are.[14]

These various hybrid breeds appear more of a side issue than a potential source of new genetic material to enrich the genome of *Felis catus*. The most promising of the species in behavioral terms, the more docile of the South American cats, are genetically incompatible with the domestic cat. The Old-World cats that are better matched genetically produce hybrids that are wilder, not calmer, than today's alley cats, so have little to offer.

Existing variation within *Felis catus* seems to be the best starting point for the completion of the cat's domestication. Plenty of modern cats combine an easygoing nature with a disinclination to hunt. Research has not yet indicated precisely how much of this variation is underpinned by genetics, but a significant proportion must be. Our goals should be to identify those individual cats with the best temperaments, and to ensure that their progeny are available to become tomorrow's pets.

One potential source of genetic variation has only recently emerged. Although house cats the world over are superficially simi-

Alley cats from the Far East (left) and Western Europe (right)

lar, their DNA reveals that they are genetically distinct—as different under the skin as, say, a Siamese and a Persian are on the surface. Interbreeding between, for example, ordinary pet cats from China with their counterparts from the UK or the United States might produce some novel temperaments, some of which might be better suited to indoor life, or be more sociable, than any of today's cats.

Selection for the right temperament among house cats requires deliberate intervention; natural evolutionary processes, which have served the cat well so far, will not be enough. One obstacle is the increasingly widespread practice of neutering cats before they breed. With so many unwanted cats euthanized every year, arguing against the widespread use of this procedure is difficult. Still, if taken too far, widespread neutering likely favors unfriendly cats over friendly. Encouraging owners not to allow their cats to produce any offspring whatsoever removes all the genes that those cats carry from being passed on to the next generation. Some of those genes have contributed to making those cats into valued pets.

When almost every pet house cat has been neutered—a situation that already applies in some parts of the UK—then we must fear for the next generations of cats. These will then mainly be the offspring of those that live on the fringes of human society—feral males, stray females, as well as those female cats owned by people who either do not care whether their cat is neutered or not, or have a moral objection to neutering. The qualities that allow most such cats to thrive and produce offspring are, unfortunately, those same behaviors we want to eliminate: wariness of people and effectiveness at self-sustaining hunting.

These cats will undoubtedly adapt their behavior to cope with whatever situations they find themselves in, but they are also likely to be genetically slightly "wilder" than the average pet cat—therefore, distinctly different from the "ideal" pet cat. Initially, the difference will probably be small, since some of the breeding cats will be strays that are genetically similar to pet cats. But as the decades pass, as fewer and fewer reproductively intact cats are available to stray, most of the kittens born each year will come from a long line of semi-wild cats—since these are the only cats that are able to breed freely. Thus, the widespread adoption of early neutering by the most responsible cat owners risks pushing the domestic cat's genetics back gradually toward the wild, away from their current domesticated state.

A study that I conducted in 1999 suggests that such extrapolation cannot be dismissed as science fiction.[15] In one area of Southampton (UK), we found that more than 98 percent of pet cat population had been neutered. So few kittens were being born that potential cat owners had to travel outside the city to obtain their cats. This situation had clearly existed for some time: from talking to the owners of the older cats, we calculated that the cat population in that area had last been self-sustaining some ten years previously, in the late 1980s. We located ten female pets in the area that were still being allowed to breed and tested the temperament of their kittens after homing, when the kittens were six months old. Our hypothesis was that feral males must have fathered many of these kittens, since so few intact males were being kept as pets in the area, and all of these were young and unlikely to compete effectively with the more wily ferals. We found that on average, the kittens in those ten litters were much less willing to settle on their owners' laps than kittens born in another area of the city that still had a significant number of undoctored pet tomcats. There was no systematic difference in the way these two groups of kittens had been socialized, and the mother cats in the two areas were indistinguishable in temperament. We therefore deduced that even if only one of the two parents comes from a long line of ferals, the kittens will be less easy to socialize than if both parents are pets. The study was too small to draw any firm conclusions, but in the years since it was carried out, blanket neutering has become more widespread, and so the cumulative effects of this on the temperament of kittens should be becoming more obvious.

Neutering is an extremely powerful selection pressure, the effects of which have been given little consideration. At present, it is the only humane way of ensuring that there are as few unwanted cats as possible, and it is unlikely ever to become so widely adopted that the house cat population begins to shrink. However, over time it will likely have unintended consequences. Consider a hypothetical situation. A century or more ago, when feline surgery was still crude, society generally accepted that most cats would reproduce. Imagine that a highly contagious parasite had appeared that sterilized cats of both sexes while they were still kittens, but otherwise left them unaffected, so that they lived as long as an unaffected cat. Any cats that happened to be resistant to that parasite would be the only cats able to breed—so, within a few years the parasite, deprived of susceptible hosts, would die out.

The only significant difference between such a hypothetical parasite and neutering is that the latter does not require a host (a cat) to continue: it lives as an idea, and so is detached from its effects.[16] Because neutering inevitably targets those cats that are being best cared for, it must logically hand the reproductive advantage to those cats that are least attached to people, many of which are genetically predisposed to remain unsocialized. We must consider the long-term effects of neutering carefully: for example, it might be better for the cats of the future as a whole if neutering programs were targeted more at ferals, which are both the unfriendliest cats and also those most likely to damage wildlife populations.

We need a fresh approach to cat breeding. Pedigree cats are bred largely for looks, not with the primary goal of ensuring an optimal temperament—although an adequate temperament is, of course, taken into account in the majority of breeds. Random-bred cats are under siege from neutering; even if this widespread practice is not making each successive generation a little wilder than the previous, it is highly unlikely to be having the opposite effect. So while cats' looks and welfare both have their champions, the cat's future has none.

Then again, why should it? Cats have always outnumbered potential owners. Why should that situation change? Cats have become more popular, not less, so there should be more homes available for them, not fewer, than there were a few decades ago. Apart from the (significant) minority of cat-haters, the general public is more tolerant of cats than of dogs. We cannot guarantee, however, that these apparent givens will continue.

Recent decades have witnessed immense changes to the way dogs are kept, especially in urban areas, with a proliferation of "clean up after your dog" regulations, no-dog parks, and legislation aimed at protecting the public from dog bites. We expect dogs to behave in a much more controlled and civilized way than they did half a century ago. Are similar restrictions on cat-keeping just around the corner? Will gardeners and wildlife enthusiasts unite to produce legislation that restricts cats to their owners' property? If such pressures do appear, they will be easier to head off if cat-lovers are already taking steps toward producing a more socially acceptable cat. At the same time, the cats themselves will benefit if they find it easier to cope with the vagaries of their pets' social lives.

Ultimately, the future of cats lies in the hands of those who breed them—not those whose eye is primarily on success in the show ring, but those who can be persuaded that an improved temperament, not good looks, should be the goal. The genetic material is available, although more science is needed to devise the temperament tests that will locate which individual cats carry it; many cats that appear well-adapted to life with people will have received an optimal upbringing, rather than being anything special genetically. The relevant genes are probably scattered all over the globe, so ideally we need collaboration between cat enthusiasts in different parts of the world.

Such human-friendly cats, however cute, are unlikely ever to command much of a price. Expectations that non-pedigree kittens should be free, or nearly so, will take a long time to die out. Commercial breeding of non-pedigree cats may never be viable. Well-adjusted kittens need a wealth of early experience that even some pedigree breeders struggle to provide. Providing this level of care is cost-effective only if kittens are bred in people's homes, the very type of environment into which they will move when they become pets.

Meanwhile, the way that cats are socialized has much room for improvement. Both breeders and owners can play a part in this, since kittens adapt to their surroundings throughout their second and third months of life. In this context, the continuing policy of some cat breeders' associations to prohibit homing until a kitten is twelve weeks of age demands careful scrutiny: it may provide extra socialization with littermates, but often at the cost of learning about different kinds of people, and the development of a robust strategy for dealing with the unfamiliar. For adult cats, training, both in the general sense of providing the right learning experiences as well as teaching them how to behave calmly in specific situations, could improve each cat's lot considerably, if only its value was more widely appreciated.

Finally, we must continue to research why some cats are strongly motivated to hunt, while the majority are content to doze in their beds. Science has not yet revealed to what extent such differences are due to early experience, and how much to genetics, but ultimately it should be possible to breed cats that are unlikely to feel the need to become predators, now that we can easily provide them with all the nutrition they need.

Cats need our understanding—both as individual animals that need our help to adjust to our ever-increasing demands, and also as a species that is still in transition between the wild and the truly domestic. If we can agree to support them in both these ways, cats will be assured a future in which they are not only popular and populous, but are also more relaxed, and affectionate, than they are today.

Further Reading

Most of my source material for this book has consisted of papers in academic journals, which are often difficult to access for those without a university affiliation. I've included references to the most important of these in the notes, with Web links if they appear to be in the public domain. For those readers who wish to take their study of cats further without first requiring a degree in biology, I can recommend the following books, most of them written by knowledgeable academics but with a more general audience in mind.

The Domestic Cat: the Biology of Its Behaviour, edited by Professors Dennis Turner and Patrick Bateson, is now available in three editions; all are published by Cambridge University Press, the most recent in 2013. These books consist of chapters written by experts in different aspects of cat behavior.

My own *The Behaviour of the Domestic Cat*, 2nd edition (Wallingford, UK: CAB International, 2012), coauthored by Drs. Sarah L. Brown and Rachel Casey, provides an integrated introduction to the science of cat behavior, aimed at an advanced undergraduate audience. *Feline Behavior: A Guide for Veterinarians* by Bonnie V. Beaver (St. Louis, MO: Saunders, 2003) is, as its title indicates, aimed at veterinary surgeons and veterinary students.

For the various stages in the history of the cat's life with humankind, Jaromir Malek's *The Cat in Ancient Egypt* (London: British Museum Press, 2006), Donald Engel's *Classical Cats* (London: Routledge,

1999), and Carl Van Vechten's *The Tiger in the House* (New York: New York Review of Books, 2006) provide specialist accounts.

Carrots and Sticks: Principles of Animal Training (Cambridge: Cambridge University Press, 2008) by Professors Paul McGreevy and Bob Boakes from the University of Sydney, Australia, is a fascinating book of two halves: the first explains learning theory in accessible language, and the second contains fifty case histories of animals (including cats) trained for specific purposes, ranging from film work to bomb detection, each illustrated with color photographs of the animals and how they were trained.

For cat owners seeking guidance on a problem cat, there is often no substitute for a one-to-one consultation with a genuine expert, but these can be hard to find. The advice given in books by Sarah Heath, Vicky Halls, or Pam Johnson-Bennett may be helpful. Celia Haddon's books may also provide some light relief.

Notes

All Web addresses mentioned in the Notes are active as of April 2013.

Introduction

1. This ratio takes many millions of unowned animals into account, and also incorporates a guess as to the numbers in Muslim countries where dogs are rare.

2. The Prophet Muhammad is said to have loved his cat Muezza so much that "he would do without his cloak rather than disturb one that was sleeping on it." Minou Reeves, *Muhammad in Europe* (New York: NYU Press, 2000), 52.

3. Rose M. Perrine and Hannah L. Osbourne, "Personality Characteristics of Dog and Cat Persons." *Anthrozoös: A Multidisciplinary Journal of the Interactions of People & Animals* 11 (1998): 33–40.

4. A recognized medical condition, referred to as "ailurophobia."

5. David A. Jessup, "The Welfare of Feral Cats and Wildlife." *Journal of the American Veterinary Medical Association* 225 (2004): 1377–83; available online at www.avma.org/News/Journals/Collections/Documents/javma_225_9_1377.pdf.

6. The People's Dispensary for Sick Animals, "The State of Our Pet Nation . . . : The PDSA Animal Wellbeing (PAW) Report 2011." Shropshire, UK: 2011; available online at tinyurl.com/b4jgzjk. Dogs scored a little better for social and physical environments (71 percent) but worse for behavior (55 percent).

7. The situation for pedigree dogs in the UK has been summarized in several expert reports, including those commissioned by the Royal Society for the Prevention of Cruelty to Animals (www.rspca.org.uk/allaboutanimals/pets/dogs/health/pedigreedogs/report), the Associate Parliamentary Group for Animal Welfare (www.apgaw.org/images/stories/PDFs/Dog-Breeding-Report-2012.pdf), and the UK Kennel Club in partnership with the re-homing charity DogsTrust (breedinginquiry.files.wordpress.com/2010/01/final-dog-inquiry-120110.pdf).

Chapter 1

1. Darcy F. Morey, *Dogs: Domestication and the Development of a Social Bond* (New York: Cambridge University Press, 2010).

2. Quoted in C. A. W. Guggisberg, *Wild Cats of the World* (New York: Taplinger, 1975), 33–34.

3. These cats are now extinct on Cyprus, displaced by the red fox, another introduction, which is now the only land-based carnivorous mammal on the island.

4. For a more detailed account of these migrations, see Stephen O'Brien and Warren Johnson's "The Evolution of Cats," *Scientific American* 297 (2007): 68–75.

5. The spelling *lybica* should more correctly be *libyca*, "from Libya," but most modern accounts use the original (incorrect) version.

6. These "lake dwellers" built villages on sites that now lie beneath the margins of lakes, but were probably fertile dry land at the time.

7. Frances Pitt (see note 9 below) claimed that the Scottish wildcat would have joined its English and Welsh counterparts in extinction, had it not been for the call-up of the younger gamekeepers to fight in the Great War.

8. Carlos A. Driscoll, Juliet Clutton-Brock, Andrew C. Kitchener, and Stephen J. O'Brien, "The Taming of the Cat," *Scientific American* 300 (2009): 68–75; available online at tinyurl.com/akxyn9c.

9. From *The Romance of Nature: Wild Life of the British Isles in Picture and Story*, vol. 2, ed. Frances Pitt (London: Country Life Press, 1936). Pitt (1888–1964) was a pioneering wildlife photographer who lived near Bridgnorth in Shropshire.

10. Mike Tomkies, *My Wilderness Wildcats* (London: Macdonald and Jane's, 1977).

11. This and the following two quotations are from Reay H. N. Smithers's "Cat of the Pharaohs: The African Wild Cat from Past to Present," *Animal Kingdom* 61 (1968): 16–23.

12. Charlotte Cameron-Beaumont, Sarah E. Lowe, and John W. S. Bradshaw, "Evidence Suggesting Preadaptation to Domestication throughout the Small Felidae," *Biological Journal of the Linnean Society* 75 (2002): 361–66; available online at www.neiu.edu/~jkasmer/Biol498R/Readings/essay1-06.pdf. In this paper, which came before Carlos Driscoll's DNA study making *cafra* a separate subspecies, the Southern African cats are listed as *Felis silvestris libyca*.

13. Carlos Driscoll et al., "The Near Eastern Origin of Cat Domestication," *Science* (Washington) 317 (2007): 519–23; available online at www.mobot.org/plantscience/resbot/Repr/Add/DomesticCat_Driscoll2007.pdf. The data discussed can be found in the online Supplemental Information, Figure S1.

14. David Macdonald, Orin Courtenay, Scott Forbes, and Paul Honess, "African Wildcats in Saudi Arabia" in *The Wild CRU Review: The Tenth Anniversary Report of the Wildlife Conservation Research Unit at Oxford University*, ed. David Macdonald and Françoise Tattersall (Oxford: University of Oxford Department of Zoology, 1996), 42.

15. The estimate of fifteen to twenty comes from Carlos Driscoll of the Laboratory of Genomic Diversity at the National Cancer Institute in Frederick, Maryland, who is currently working to pinpoint where these genes lie on the cat's chromosomes, and how they may work.

16. See note 13 above.

17. O. Bar-Yosef, "Pleistocene Connexions between Africa and Southwest Asia: An Archaeological Perspective," *The African Archaeological Review* 5 (1987), 29–38.

18. Carlos Driscoll and his colleagues have discovered five distinct types of mitochondrial DNA in today's domestic cats; mitochondrial DNA is inherited only through the maternal line. The common maternal ancestor of these five cats lived about 130,000 years ago; over the next 120,000 years, her descendants gradually moved around the Middle East and North Africa, their mitochondrial DNA mutating slightly from time, before a few of them happened to become the ancestors of today's pet cats.

Chapter 2

1. J.-D. Vigne, J. Guilane, K. Debue, L. Haye, and P. Gérard, "Early Taming of the Cat in Cyprus," *Science* 304 (2004): 259.

2. James Serpell, *In the Company of Animals: A Study of Human-Animal Relationships*, Canto ed. (New York: Cambridge University Press, 1996); Stefan Seitz, "Game, Pets and Animal Husbandry among Penan and Punan Groups" in *Beyond the Green Myth: Borneo's Hunter-Gatherers in the Twenty-first Century*, ed. Peter G. Sercombe and Bernard Sellato (Copenhagen: NIAS Press, 2007).

3. Veerle Linseele, Wim Van Neer, and Stan Hendrickx, "Evidence for Early Cat Taming in Egypt," *Journal of Archaeological Science* 34 (2007): 2081–90 and 35 (2008): 2672–73; available online at tinyurl.com/aotk2e8.

4. Jaromír Málek, *The Cat in Ancient Egypt* (London, British Museum Press, 2006).

5. This stone coffin is now in the Museum of Egyptian Antiquities in Cairo. On its side, alongside pictures of Ta-Miaut herself and the deities Nephthys and Isis, are several inscriptions. Words spoken by Osiris: "Ta-Miaut is not tired, nor weary is the body of Ta-Miaut, justified before the Great God." Words spoken by Isis: "I embrace you in my arms, Osiris." Words spoken by Nephthys: "I envelop my brother, Osiris Ta-Miaut, the Triumphant." See www.mafdet.org/tA-miaut.html.

6. At the same time, cats may have been implicated in the first outbreaks of bubonic plague. Although this disease was later spread into Europe by black rats, its natural host is apparently the Nile rat. Although the disease is usually transmitted from the Nile rat to humans via the rat flea, cat fleas are also sometimes responsible. See Eva Panagiotakopulu, "Pharaonic Egypt and the Origins of Plague," *Journal of Biogeography* 31 (2004): 269–75; available online at tinyurl.com/ba52zuv.

7. Both genets and mongooses are occasionally kept as domestic pets, but these are genetically unaltered from their wild ancestors, not domesticated animals, and consequently difficult to keep.

8. From *The Historical Library of Diodorus the Sicilian*, Vol. 1, Chap. VI, trans. G. Booth (London: Military Chronicle Office, 1814), 87.

9. Frank J. Yurko, "The Cat and Ancient Egypt," *Field Museum of Natural History Bulletin* 61 (March–April 1990): 15–23.

10. Mongooses have been introduced to many parts of the world in an attempt to control snakes, especially islands such as Hawaii and Fiji, which lack other predators of snakes.

11. Angela von den Driesch and Joachim Boessneck, "A Roman Cat Skeleton from Quseir on the Red Sea Coast," *Journal of Archaeological Science* 10 (1983): 205–11.

12. Herodotus, *The Histories (Euterpe)* 2:60, trans. G. C. Macaulay (London & New York: MacMillan & Co., 1890).

13. Herodotus, *Histories*, 2:66.

14. From *The Historical Library of Diodorus the Sicilian*, Vol. 1, Chap. VI, 84, trans. G. Booth (London: Military Chronicle Office, 1814), 84.

15. Herodotus, *Histories (Euterpe)*, 2:66.

16. Elizabeth Marshall Thomas, *The Tribe of Tiger: Cats and Their Culture* (New York: Simon & Schuster, 1994), 100–01.

17. Paul Armitage and Juliet Clutton-Brock, "A Radiological and Histological Investigation into the Mummification of Cats from Ancient Egypt," *Journal of Archaeological Science* 8 (1981): 185–96.

18. Stephen Buckley, Katherine Clark, and Richard Evershed, "Complex Organic Chemical Balms of Pharaonic Animal Mummies," *Nature* 431 (2004): 294–99.

19. Armitage and Clutton-Brock.

20. The "black panther," a melanistic form of the leopard, is common in the rainforests of southern Asia. Presumably so little light penetrates to the forest floor that camouflage is not as much of an issue as for a normally spotted leopard hunting in the African bush.

21. Neil B. Todd, who collected this data, instead suggests that the orange mutation first arose in Asia Minor (roughly, modern Turkey), even though it is less common there than in Alexandria. "Cats and Commerce," *Scientific American* 237 (1977): 100–07.

22. Dominique Pontier, Nathalie Rioux, and Annie Heizmann, "Evidence of Selection on the Orange Allele in the Domestic Cat *Felis catus*: The Role of Social Structure." *Oikos* 73 (1995): 299–308.

23. Terence Morrison-Scott, "The Mummified Cats of Ancient Egypt," *Proceedings of the Zoological Society of London* 121 (1952): 861–67.

24. See chapter 16 of Frederick Everard Zeuner's *A History of Domesticated Animals* (New York: Harper & Row, 1963).

25. This rapid acceptance of cats contrasts with the Japanese refusal to admit dogs for thousands of years after they had become widely adopted in China.

26. Monika Lipinski et al., "The Ascent of Cat Breeds: Genetic Evaluations of Breeds and Worldwide Random-bred Populations," *Genomics* 91 (2008): 12–21; available online at tinyurl.com/cdop2op.

27. Cleia Detry, Nuno Bicho, Hermenegildo Fernandes, and Carlos Fernandes, "The Emirate of Córdoba (756–929 AD) and the Introduction of the Egyptian Mongoose *(Herpestes ichneumon)* in Iberia: The Remains from Muge, Portugal," *Journal of Archaeological Science* 38 (2011): 3518–23. The related Indian mongoose has been introduced into many parts of the world in an attempt to control snakes, especially islands such as Hawaii and Fiji, which lack other snake predators.

28. Lyudmila N. Trut, "Early Canid Domestication: The Farm-Fox Experiment," *American Scientist* 87 (1999): 160–69; available online at www.terrierman.com/russianfoxfarmstudy.pdf.

Chapter 3

1. Perhaps fortunately for the domestic cat's popularity, the brown rat *Rattus norvegicus*, much larger and formidable than the black rat, did not spread across Europe until the late Middle Ages. As it advanced, it gradually displaced the plague-carrying black rat from the towns and cities; now, black rats are generally found only in warmer places. Most cats are not powerful or skillful enough to take on a full-grown brown rat, although they can be an effective deterrent to brown rat colonization or recolonization. See Charles Elton, "The Use of Cats in Farm Rat Control," *British Journal of Animal Behaviour* 1 (1953), 151–55.

2. Researchers have found numerous examples of this in Britain, France, and Spain, so this superstition must have been widespread. A mummified cat/rat pair were even discovered in Dublin's Christ Church Cathedral, although the official story holds that they were trapped there by accident.

3. Translation by Eavan Boland; see homepages.wmich.edu/~cooneys/poems/pangur.ban.html. Bán means "white" in Old Irish, so presumably that was the color of the writer's cat.

4. Ronald L. Ecker and Eugene J. Crook, *Geoffrey Chaucer:* The Canterbury Tales—*A Complete Translation into Modern English* (Online Edition; Palatka, FL: Hodge & Braddock, 1993); english.fsu.edu/canterbury.

5. Tom P. O'Connor, "Wild or Domestic? Biometric Variation in the Cat *Felis silvestris Schreber*," *International Journal of Osteoarchaeology* 17 (2007): 581–95; available online at eprints.whiterose.ac.uk/3700/1/OConnor_Cats-IJOA-submitted.pdf.

6. At this time, cats were also considered generally detrimental to good health. The French physician Ambroise Paré described the cat as "a venomous animal which infects through its hair, its breath and its brains," and in 1607 the English cleric Edward Topsell wrote, "It is most certain that the breath and savour of cats . . . destroy the lungs."

7. Carl Van Vechten, *The Tiger in the House*, 3rd ed. (New York: Dorset Press, 1952), 100.

8. J. S. Barr: Buffon's *Natural History*, Vol. VI, translated from the French (1797), 1.

9. Neil Todd, "Cats and Commerce," *Scientific American* 237 (May 1977): 100–07.

10. Ibid.

11. Manuel Ruiz-García and Diana Alvarez, "A Biogeographical Population Genetics Perspective of the Colonization of Cats in Latin America and Temporal Genetic Changes in Brazilian Cat Populations," *Genetics and Molecular Biology* 31 (2008): 772–82.

12. Even black cats carry the genes for one or other of the "tabby" patterns, but because the hairs that should be brown at the tips are black, the pattern doesn't show—at least, not until the cat gets old, when the hairs that would be brown if they weren't black fade to a dark, rusty color. A tabby pattern can also be just discernible in black kittens for a few weeks. Another variation of the tabby gene, "Abyssinian," restricts the black stripes to the head, tail, and legs, while the body is

covered in brown-tipped hairs; this is quite rare except in the pedigree cats of the same name.

13. See Todd, footnote 9.

14. Bennett Blumenberg, "Historical Population Genetics of *Felis catus* in Humboldt County, California," *Genetica* 68 (1986): 81–86.

15. Andrew T. Lloyd, "Pussy Cat, Pussy Cat, Where Have You Been?" *Natural History* 95 (1986): 46–53.

16. Ruiz-García and Alvarez, note 11.

17. Manuel Ruiz-García, "Is There Really Natural Selection Affecting the L Frequencies (Long Hair) in the Brazilian Cat Populations?" *Journal of Heredity* 91 (2000): 49–57.

18. Juliet Clutton-Brock, formerly of the British Museum of Natural History in London, pointed this out in her 1987 book *A Natural History of Domesticated Mammals* (New York: Cambridge University Press). Indian elephants, camels, and reindeer are among other domestic animals that exist, like the domestic cat, somewhere between wildness and full domestication.

19. For more detail on cat nutrition and how it interacts with their lifestyles, see Debra L. Zoran and C. A. T. Buffington, "Effects of Nutrition Choices and Lifestyle Changes on the Well-being of Cats, a Carnivore that Has Moved Indoors," *Journal of the American Veterinary Medical Association* 239 (2011): 596–606.

20. The idea of "nutritional wisdom" comes from Chicago-based pediatrician Clara Marie Davis's classic 1933 experiment, which showed that human infants, allowed to choose from thirty-three "natural" foodstuffs, would choose a balanced diet, even though each infant preferred a different combination of foods.

21. Stuart C. Church, John A. Allen and John W. S. Bradshaw, "Frequency-Dependent Food Selection by Domestic Cats: A Comparative Study," *Ethology* 102 (1996): 495–509.

Chapter 4

1. I suspect—and this is only conjecture—that it is no coincidence that both cat and dog are members of the Carnivora.

2. Eileen Karsh, *The Domestic Cat: The Biology of Its Behavior*, 2nd ed., Dennis C. Turner and Patrick Bateson (New York: Cambridge University Press, 1988), 164. Professor Karsh and her team carried out their study at Temple University in Philadelphia. Remarkably, this revolutionary work has never been published in peer-reviewed journals; however, no one since has fundamentally disagreed with its conclusions.

3. M. E. Pozza, J. L. Stella, A.-C. Chappuis-Gagnon, S. O. Wagner, and C. A. T. Buffington, "Pinch-induced Behavioral Inhibition ('Clipnosis') in Domestic Cats," *Journal of Feline Medicine and Surgery* 10 (2008): 82–87.

4. John M. Deag, Aubrey Manning, and Candace E. Lawrence, "Factors Influencing the Mother-Kitten Relationship," in *The Domestic Cat*, 23–39.

5. Jay S. Rosenblatt, "Suckling and Home Orientation in the Kitten: A Comparative Developmental Study," in *The Biopsychology of Development*, ed. Ethel Tobach, Lester R. Aronson, and Evelyn Shaw (New York: Academic Press, 1971), 345–410.

6. R. Hudson, G. Raihani, D. González, A. Bautista, and H. Distel, "Nipple Preference and Contests in Suckling Kittens of the Domestic Cat Are Unrelated to Presumed Nipple Quality," *Developmental Psychobiology* 51 (2009): 322–32.

7. St. Francis Animal Welfare in Fair Oak, Hampshire, UK.

8. Female cats sometimes mate with several males in succession, such that the members of a litter may be half-siblings. See the chapter by Olof Liberg, Mikael Sandell, Dominique Pontier, and Eugenia Natoli in *The Domestic Cat*, 119–147.

9. Hand-reared kittens often end up spending their entire lives with the person that hand-reared them. Whether this is because the kittens are difficult to home or whether their human foster parents cannot bear to give them away seems to be unknown.

10. John Bradshaw and Suzanne L. Hall, "Affiliative Behaviour of Related and Unrelated Pairs of Cats in Catteries: A Preliminary Report," *Applied Animal Behaviour Science* 63 (1999): 251–55.

11. Roberta R. Collard, "Fear of Strangers and Play Behavior in Kittens with Varied Social Experience," *Child Development* 38 (1967): 877–91.

12. Karsh, *The Domestic Cat*, note 2.

13. We measured the closeness of the relationship by asking the owners how likely they would be to turn to their cat for emotional support in each of nine situations—for example, after a bad day at work, or when they were feeling lonely. See Rachel A. Casey and John Bradshaw, "The Effects of Additional Socialisation for Kittens in a Rescue Centre on Their Behaviour and Suitability as a Pet," *Applied Animal Behaviour Science* 114 (2008): 196–205.

14. Practical steps for minimizing the stress of being moved to a new house—for both kittens and cats—can be found on the Cats Protection website: www.cats.org.uk/uploads/documents/cat_care_leaflets/EG02-Welcomehome.pdf.

15. Although purring is not always a reliable indicator of how friendly a kitten is, it probably was in this instance.

16. Perhaps surprisingly, the "owners" of these half-wild cats didn't seem to mind—some people seem to value cats for their wildness, and may even deliberately choose a cat with a personality to match.

Chapter 5

1. Birds, much more visual creatures than cats, see four colors, including ultraviolet, invisible to mammals of all kinds.

2. At least we know that people who are red-green color-blind see colors this way. A very small number of people also have one normal eye and one red-green color-blind eye, and can develop a normal vocabulary for color using their "good" eye, and then using that vocabulary to report what they see with only their color-blind eye open.

3. You can try this by placing a finger on the page of this book, and then moving the finger a little way toward your nose while continuing to focus on the print. We can choose to fix our gaze on either the print or the finger, but if we had a cat's eyes, at this distance, we would be unable to fix on the finger.

4. David McVea and Keir Pearson, "Stepping of the Forelegs over Obstacles Establishes Long-lasting Memories in Cats," *Current Biology* 17 (2007): R621–23.

5. See also the animation at en.wikipedia.org/wiki/Cat_righting_reflex.

6. Nelika K. Hughes, Catherine J. Price, and Peter B. Banks, "Predators Are Attracted to the Olfactory Signals of Prey," *PLoS One* 5 (2010): e13114, doi: 10.1371.

7. A vestigial Jacobson's organ can be detected in the human fetus, but it never develops functional nerve connections.

8. Ignacio Salazar, Pablo Sanchez Quinteiro, Jose Manuel Cifuentes, and Tomas Garcia Caballero, "The Vomeronasal Organ of the Cat," *Journal of Anatomy* 188 (1996): 445–54.

9. Whereas mammals seem to use their VNOs exclusively for social and especially sexual functions, reptiles use them more diversely. Snakes use their forked tongues, which do not have taste buds, to deliver different samples of odorants to their left and right VNOs, useful when they are tracking prey or a snake of the opposite sex.

10. See Patrick Pageat and Emmanuel Gaultier, "Current Research in Canine and Feline Pheromones," *The Veterinary Clinics: North American Small Animal Practice* 33 (2003): 187–211.

Chapter 6

1. However, science has recently revealed that some of our emotions never surface into consciousness but nevertheless affect the way we behave; for example, images and emotions that we never become aware of affect the way we drive our cars. See Ben Lewis-Evans, Dick de Waard, Jacob Jolij, and Karel A. Brookhuis, "What You May Not See Might Slow You Down Anyway: Masked Images and Driving," *PLoS One* 7 (2012): e29857, doi: 10.1371/journal.pone.0029857.

2. Leonard Trelawny Hobhouse, *Mind in Evolution*, 2nd ed. (London: Macmillan & Co., 1915).

3. For details of the experiments, see M. Bravo, R. Blake, and S. Morrison, "Cats See Subjective Contours," *Vision Research* 18 (1988): 861–65; F. Wilkinson, "Visual Texture Segmentation in Cats," *Behavioural Brain Research* 19 (1986): 71–82.

4. Further details of the discriminatory abilities of cats can be found in John W. S. Bradshaw, Rachel A. Casey, and Sarah L. Brown, *The Behaviour of the Domestic Cat*, 2nd ed. (Wallingford, UK: CAB International, 2012), chap. 3.

5. Sarah L. Hall, John W. S. Bradshaw, and Ian Robinson, "Object Play in Adult Domestic Cats: The Roles of Habituation and Disinhibition," *Applied Animal Behaviour Science* 79 (2002): 263–71. Compared to "classic" habituation as studied in laboratory rats, the timescale over which cats remain habituated to toys is very long—minutes, rather than seconds. We subsequently found that the same applies to dogs.

6. Sarah L. Hall and John W. S. Bradshaw, "The Influence of Hunger on Object Play by Adult Domestic Cats," *Applied Animal Behaviour Science* 58 (1998): 143–50.

7. Commercially available toys don't come apart for a good reason: occasionally a cat can choke on a piece of toy, or get fragments lodged in its gut.

8. Although cats will go out hunting whether they're hungry or not, they're more likely to make a kill if they're hungry; see chapter 10.

9. For a (very) alternative view, look online for comedian Eddie Izzard's "Pavlov's Cat," currently at tinyurl.com/dce6lb.

10. Psychologists generally classify pain as a feeling rather than an emotion, but there is no doubt that both feelings and emotions are equally involved in how animals learn about the world.

11. Endre Grastyán and Lajos Vereczkei, "Effects of Spatial Separation of the Conditioned Signal from the Reinforcement: A Demonstration of the Conditioned Character of the Orienting Response or the Orientational Character of Conditioning," *Behavioral Biology* 10 (1974): 121–46.

12. Ádam Miklósi, Péter Pongrácz, Gabriella Lakatos, József Topál, and Vilmos Csányi, "A Comparative Study of the Use of Visual Communicative Signals in Interactions Between Dogs (*Canis familiaris*) and Humans and Cats (*Felis catus*) and Humans," *Journal of Comparative Psychology* 119 (2005): 179–86; available online at www.mtapi.hu/userdirs/26/Publikaciok_Topal/Miklosietal2005JCP.pdf.

13. Nicholas Nicastro and Michael J. Owren, "Classification of Domestic Cat (*Felis catus*) Vocalizations by Naive and Experienced Human Listeners," *Journal of Comparative Psychology* 117 (2003): 44–52.

14. Edward L. Thorndike, *Animal Intelligence: An Experimental Study of the Associative Processes in Animals*, Chapter 2 (New York: Columbia University Press, 1898); available online at tinyurl.com/c4bl6do.

15. Emma Whitt, Marie Douglas, Britta Osthaus, and Ian Hocking, "Domestic Cats (*Felis catus*) Do Not Show Causal Understanding in a String-pulling Task," *Animal Cognition* 12 (2009): 739–43. The same scientists had earlier shown that the crossed-strings arrangement defeats most dogs, even those which, unlike the cats, had earlier successfully solved the parallel-strings problem—thus, dogs' understanding of physics seems to be better than cats'.

16. Claude Dumas, "Object Permanence in Cats (*Felis catus*): An Ecological Approach to the Study of Invisible Displacements," *Journal of Comparative Psychology* 106 (1992): 404–10; Dumas, "Flexible Search Behavior in Domestic Cats (*Felis catus*): A Case Study of Predator-Prey Interaction," *Journal of Comparative Psychology* 114 (2000): 232–38.

17. For other works by this cartoonist, visit www.stevenappleby.com.

18. George S. Romanes, *Animal Intelligence* (New York: D. Appleton & Co., 1886); available online at www.gutenberg.org/files/40459/40459-h/40459-h.htm.

19. C. Lloyd Morgan, *An Introduction to Comparative Psychology* (New York: Scribner, 1896); available online at tinyurl.com/crehpj9.

20. Paul H. Morris, Christine Doe, and Emma Godsell, "Secondary Emotions in Non-primate Species? Behavioural Reports and Subjective Claims by Animal Owners," *Cognition and Emotion* 22 (2008): 3–20.

21. More detail of the causes of such problematic behavior can be found in chapters 11 and 12 of my book, *The Behaviour of the Domestic Cat*, 2nd ed., coauthored with Rachel Casey and Sarah Brown (Wallingford, UK: CAB International, 2012).

22. Anne Seawright et al., "A Case of Recurrent Feline Idiopathic Cystitis: The Control of Clinical Signs with Behavior Therapy," *Journal of Veterinary Behavior: Clinical Applications and Research* 3 (2008): 32–38. For background information on

feline cystitis, see the Feline Advisory Bureau's website, www.fabcats.org/owners/flutd/info.html.

23. Alexandra Horowitz, a professor of cognitive psychology at New York's Barnard College, performed this study. See her paper "Disambiguating the 'Guilty Look': Salient Prompts to a Familiar Dog Behaviour" in *Behavioural Processes* 81 (2009): 447–52, and her book *Inside of a Dog: What Dogs See, Smell, and Know* (New York: Scribner, 2009).

Chapter 7

1. "The Curious Cat," filmed for the BBC's *The World About Us* series (1979). A delightful account of the making of this film is included in its companion volume of the same title, written by Michael Allaby and Peter Crawford (London: M. Joseph, 1982). Similar studies were being conducted at roughly the same time by Jane Dards in Portsmouth dockyard (UK), Olof Liberg in Sweden, and Masako Izawa in Japan.

2. Strictly speaking, the term "gene" refers to a single location on a particular chromosome, and the competing versions of the same gene are *alleles*; one example already discussed is the blotched and striped alleles of the "tabby" gene.

3. This is likely true even though we know little about how such genes might work, since almost all cooperation between animals occurs between members of the same family. Genes code for proteins, and it is difficult to imagine a protein that could promote family loyalty as such. Rather, many genes must be involved, each contributing a small piece to the whole: for example, one might increase the threshold for aggression toward other cats in general, while another enables the recognition of odors characteristic of family members, through changes in the accessory olfactory bulb, the part of the brain that processes information coming from the vomeronasal organ.

4. Christopher N. Johnson, Joanne L. Isaac, and Diana O. Fisher, "Rarity of a Top Predator Triggers Continent-wide Collapse of Mammal Prey: Dingoes and Marsupials in Australia," *Proceedings of the Royal Society B* 274 (2007): 341–46.

5. Dominique Pontier and Eugenia Natoli, "Infanticide in Rural Male Cats (*Felis catus* L.) as a Reproductive Mating Tactic," *Aggressive Behavior* 25 (1999): 445–49.

6. Phyllis Chesler, "Maternal Influence in Learning by Observation in Kittens," *Science* 166 (1969): 901–03.

7. Marvin J. Herbert and Charles M. Harsh, "Observational Learning by Cats," *Journal of Comparative Psychology* 31 (1944): 81–95.

8. Transcribed from the original letter in the British Museum (Natural History) collections.

9. These studies were mainly conducted by my colleagues Sarah Brown and Charlotte Cameron Beaumont. For more detail, see chapter 8 of my book, *The Behaviour of the Domestic Cat*, 2nd ed., coauthored by Rachel Casey and Sarah Brown (Wallingford, UK: CAB International, 2012).

10. This silhouette trick fooled most cats only once; the second one they came across elicited almost no reaction at all.

11. Researchers think that the retention of juvenile characteristics into adulthood, referred to as neoteny, was a major factor in the domestication of many animals, especially the dog. For example, at first sight the skull of a Pekinese is nothing like that of its ancestor the wolf, but in fact it is roughly the same shape as the skull of a wolf fetus. Although the domestic cat's body has not been neotenized, some of its behavior may have been, including the upright tail and some other social signals.

12. A fuller account of the evolution of signaling in the cat family can be found in John W. S. Bradshaw and Charlotte Cameron-Beaumont, "The Signalling Repertoire of the Domestic Cat and Its Undomesticated Relatives," in Dennis Turner and Patrick Bateson, eds., *The Domestic Cat: The Biology of Its Behaviour*, 2nd ed. (Cambridge: Cambridge: Cambridge University Press, 2000), 67–93.

13. Christina D. Buesching, Pavel Stopka, and David W. Macdonald, "The Social Function of Allo-marking in the European Badger (*Meles meles*)," *Behaviour* 140 (2003): 965–80.

14. Terry Marie Curtis, Rebecca Knowles, and Sharon Crowell-Davis, "Influence of Familiarity and Relatedness on Proximity and Allogrooming in Domestic Cats (*Felis catus*)," *American Journal of Veterinary Research* 64 (2003): 1151–54. See also Ruud van den Bos, "The Function of Allogrooming in Domestic Cats (*Felis silvestris catus*): A Study in a Group of Cats Living in Confinement," *Journal of Ethology* 16 (1998): 1–13.

15. See my book *Dog Sense: How the New Science of Dog Behavior Can Make You a Better Friend to Your Pet* (New York: Basic Books, 2011).

16. Feral dogs, the direct descendants of wolves, show little of their ancestors' social sophistication. Although groups of males and females band together to form packs that share common territory, all the adult females attempt to breed, and most of the puppies are cared for only by their mother—although occasionally, two litters may be pooled, and some records show fathers bringing food to their litters.

17. One well-researched instance of explosive speciation is the cichlid fish of Lake Victoria, which despite now being the world's largest tropical lake was dry land only 15,000 years ago. Today it contains many hundreds of cichlid species, none of which are found in any of the other African great lakes, and most of which have evolved in the 14,000 years since the lake last filled with water. See, for example, Walter Salzburger, Tanja Mack, Erik Verheyen, and Axel Meyer, "Out of Tanganyika: Genesis, Explosive Speciation, Key-innovations and Phylogeography of the Haplochromine Cichlid Fishes," *BMC Evolutionary Biology* 5 (2005): 17.

18. Kipling, *Just So Stories for Little Children* (New York: Doubleday, Page & Co., 1902); available online at www.boop.org/jan/justso/cat.htm.

19. Our noses are particularly sensitive to thiols: minute traces are added to natural gas, which in itself is completely odorless, to make it easier to detect leaks.

20. Ludovic Say and Dominique Pontier, "Spacing Pattern in a Social Group of Stray Cats: Effects on Male Reproductive Success," *Animal Behaviour* 68 (2004): 175–80.

21. See, for example, the policies of Cats Protection, which is as of 2011 "the largest single cat neutering group in the world": www.cats.org.uk/what-we-do/neutering/.

Chapter 8

1. Gary D. Sherman, Jonathan Haidt, and James A. Coan, "Viewing Cute Images Increases Behavioral Carefulness," *Emotion* 9 (2009): 282–86; available online at tinyurl.com/bxqg2u6.

2. Robert A. Hinde and Les A. Barden, "The Evolution of the Teddy Bear," *Animal Behaviour* 33 (1985): 1371–73.

3. See www.wwf.org.uk/how_you_can_help/the_panda_made_me_do_it/.

4. Kathy Carlstead, Janine L. Brown, Steven L. Monfort, Richard Killens, and David E. Wildt, "Urinary Monitoring of Adrenal Responses to Psychological Stressors in Domestic and Nondomestic Felids," *Zoo Biology* 11 (1992): 165–76.

5. Susan Soennichsen and Arnold S. Chamove, "Responses of Cats to Petting by Humans," *Anthrozoös: A Multidisciplinary Journal of the Interactions of People & Animals* 15 (2002): 258–65.

6. Karen McComb, Anna M. Taylor, Christian Wilson, and Benjamin D. Charlton, "The Cry Embedded within the Purr," *Current Biology* 19 (2009): R507–08.

7. Henry W. Fisher, *Abroad with Mark Twain and Eugene Field: Tales They Told to a Fellow Correspondent* (New York: Nicolas L. Brown, 1922), 102. It's worth noting the end of the quotation as well: " . . . outside of the girl you love, of course."

8. Some manufacturers add salt to dry cat food, but not for its taste: it's mainly there to stimulate cats to drink, thereby minimizing their risk of developing bladder stones.

9. The late Penny Bernstein conducted a detailed study of stroking, details of which sadly remained unpublished when she died in 2012. For a summary, see Tracy Vogel's "Petting Your Cat—Something to Purr About" at www.pets.ca/cats/articles/petting-a-cat/, and Bernstein's own review, "The Human-Cat Relationship," in *The Welfare of Cats*, ed. Irene Rochlitz (Dordrecht, the Netherlands: Springer, 2005), 47–89.

10. Soennichsen and Chamove, "Responses of Cats," note 5.

11. Mary Louise Howden, "Mark Twain as His Secretary at Stormfield Remembers Him: Anecdotes of the Author Untold until Now," *New York Herald*, December 13, 1925, 1–4; available online at www.twainquotes.com/howden.html.

12. Sarah Lowe and John W. S. Bradshaw, "Ontogeny of Individuality in the Domestic Cat in the Home Environment," *Animal Behaviour* 61 (2001): 231–37.

13. To hear two Bengal cats yowling and chirruping to each other, go to tinyurl.com/crb5ycj. The noise that many cats make when they see birds through a window is sometimes referred to as "chirping," but is more correctly "chattering"; see tinyurl.com/cny83rd.

14. Mildred Moelk, "Vocalizing in the House-cat: A Phonetic and Functional Study," *American Journal of Psychology* 57 (1944): 184–205.

15. Nicholas Nicastro, "Perceptual and Acoustic Evidence for Species-level Differences in Meow Vocalizations by Domestic Cats (*Felis catus*) and African Wild Cats (*Felis silvestris lybica*)," *Journal of Comparative Psychology* 118 (2004): 287–96. When this paper was published, it was common practice to refer to all African wild-

cats as *lybica*; however, these southern African wildcats, now known as *cafra*, are not particularly closely related to domestic cats, having diverged from the Middle-Eastern/North-African *lybica* more than 150,000 years ago.

16. Nicholas Nicastro and Michael J. Owren, "Classification of Domestic Cat (*Felis catus*) Vocalizations by Naive and Experienced Human Listeners," *Journal of Comparative Psychology* 117 (2003): 44–52.

17. Dennis C. Turner, "The Ethology of the Human-Cat Relationship," *Swiss Archive for Veterinary Medicine* 133 (1991): 63–70.

18. Desmond Morris, *Catwatching: Why Cats Purr and Other Feline Mysteries Explained* (New York: Three Rivers Press, 1993).

19. Of course, many cats do groom their owners, but cats also groom other adult cats.

20. Maggie Lilith, Michael Calver, and Mark Garkaklis, "Roaming Habits of Pet Cats on the Suburban Fringe in Perth, Western Australia: What Size Buffer Zone Is Needed to Protect Wildlife in Reserves?" in Daniel Lunney, Adam Munn, and Will Meikle, eds., *Too Close for Comfort: Contentious Issues in Human-Wildlife Encounters* (Mosman, NSW, Australia: Royal Zoological Society of New South Wales, 2008), 65–72. See also Roland W. Kays and Amielle A. DeWan, "Ecological Impact of Inside/Outside House Cats around a Suburban Nature Preserve," *Animal Conservation* 7 (2004): 273–83; available online at www.nysm.nysed.gov/staffpubs/docs/15128.pdf.

21. The transmitter and its battery could be carried by a bird the size of a thrush, so it was extremely light. These radio collars emit "beeps" every few seconds that are picked up on a portable antenna and receiver; the aerial is directional, producing the strongest signal when pointed directly at the animal. When used for tracking wildlife, the operator keeps at a distance once a reasonably strong signal has been picked up, to avoid disturbing the animal, taking several recordings from different angles to map the animal's exact location. With a pet cat, it's easier to simply walk toward the radio source until the cat is sighted.

22. This feline unfaithfulness was exposed by the University of Georgia's Kitty Cam project: fifty-five cats in Athens, Georgia wore lightweight video recorders, revealing that four often visited second households where they received food and/or affection. See the website for *The National Geographic & University of Georgia Kitty Cams (Crittercam) Project: A Window into the World of Free-roaming Cats* (2011), www.kittycams.uga.edu/research.html.

23. Adapted from material provided by Rachel Casey. See John W. S. Bradshaw, Rachel Casey, and Sarah Brown, *The Behaviour of the Domestic Cat*, 2nd ed. (Wallingford, UK: CAB International, 2012), chapter 11.

24. Ibid.

25. Ronald R. Swaisgood and David J. Shepherdson, "Scientific Approaches to Enrichment and Stereotypes in Zoo Animals: What's Been Done and Where Should We Go Next?" *Zoo Biology* 24 (2005): 499–518.

26. Marianne Hartmann-Furter, "A Species-Specific Feeding Technique Designed for European Wildcats (*Felis s. silvestris*) in Captivity," *Säugetierkundliche Informationen* 4 (2000): 567–75.

27. Adapted from the RSPCA webpage "Keeping Cats Indoors" (2013), www.rspca.org.uk/allaboutanimals/pets/cats/environment/indoors.

28. See Cat Behavior Associates, *The Benefits of Using Puzzle Feeders for Cats* (2013), www.catbehaviorassociates.com/the-benefits-of-using-puzzle-feeders-for-cats/.

Chapter 9

1. Monika Lipinski et al., "The Ascent of Cat Breeds: Genetic Evaluations of Breeds and Worldwide Random-bred Populations," *Genomics* 91 (2008): 12–21.

2. See www.gccfcats.org/breeds/oci.html.

3. Bjarne O. Braastad, I. Westbye, and Morten Bakken, "Frequencies of Behaviour Problems and Heritability of Behaviour Traits in Breeds of Domestic Cat," in Knut Bøe, Morten Bakken, and Bjarne Braastad, eds., *Proceedings of the 33rd International Congress of the International Society for Applied Ethology, Lillehammer, Norway* (Ås: Agricultural University of Norway, 1999), 85.

4. Paola Marchei et al., "Breed Differences in Behavioural Response to Challenging Situations in Kittens," *Physiology & Behavior* 102 (2011): 276–84.

5. See chapter 11 of my book *Dog Sense* (New York: Basic Books, 2011).

6. Obviously, the mother cat makes genetic contributions to the kittens as well, but she can also influence the development of her kittens' personalities through the way she raises them. Thus, the effects of maternal genetics, while undoubtedly as strong as those of the father, are harder to pin down.

7. For a more detailed discussion, see Sarah Hartwell, "Is Coat Colour Linked to Temperament?" (2001), www.messybeast.com/colour-tempment.htm.

8. Michael Mendl and Robert Harcourt, "Individuality in the Domestic Cat: Origins, Development and Stability," in *The Domestic Cat*, 47–64.

9. Rebecca Ledger and Valerie O'Farrell, "Factors Influencing the Reactions of Cats to Humans and Novel Objects," in Ian Duncan, Tina Widowski, and Derek Haley, eds., *Proceedings of the 30th International Congress of the International Society for Applied Ethology* (Guelph: Col. K. L. Campbell Centre for the Study of Animal Welfare, 1996), 112.

10. Caroline Geigy, Silvia Heid, Frank Steffen, Kristen Danielson, André Jaggy, and Claude Gaillard, "Does a Pleiotropic Gene Explain Deafness and Blue Irises in White Cats?" *The Veterinary Journal* 173 (2007): 548–53.

11. John W. S. Bradshaw and Sarah Cook, "Patterns of Pet Cat Behaviour at Feeding Occasions," *Applied Animal Behaviour Science* 47 (1996): 61–74.

12. For a review, see Michael Mendl and Robert Harcourt, note 8.

13. Sandra McCune, "The Impact of Paternity and Early Socialisation on the Development of Cats' Behaviour to People and Novel Objects," *Applied Animal Behaviour Science* 45 (1995): 109–24.

14. Sarah E. Lowe and John W. S. Bradshaw, "Responses of Pet Cats to Being Held by an Unfamiliar Person, from Weaning to Three Years of Age," *Anthrozoös* 15 (2002): 69–79.

15. The technical term for an inveterate cat-hater is an ailurophobe. I carried out my research here with the assistance of a former colleague, Dr. Deborah Goodwin.

16. Kurt Kotrschal, Jon Day, and Manuela Wedl, "Human and Cat Personalities: Putting Them Together," in Dennis C. Turner and Patrick Bateson, eds., *The Domestic Cat: The Biology of Its Behaviour*, 3rd ed. (Cambridge: Cambridge University Press, 2014).

17. Jill Mellen, "Effects of Early Rearing Experience on Subsequent Adult Sexual Behavior Using Domestic Cats (*Felis catus*) as a Model for Exotic Small Felids," *Zoo Biology* 11 (1992): 17–32.

18. Nigel Langham, "Feral Cats (*Felis catus* L.) on New Zealand Farmland. II. Seasonal Activity," *Wildlife Research* 19 (1992): 707–20.

19. Julie Feaver, Michael Mendl, and Patrick Bateson, "A Method for Rating the Individual Distinctiveness of Domestic Cats," *Animal Behaviour* 34 (1986): 1016–25.

20. See Patrick Bateson, "Behavioural Development in the Cat," in Turner and Bateson, eds., *The Domestic Cat*, 2nd ed., 9–22.

Chapter 10

1. Michael C. Calver, Jacky Grayson, Maggie Lilith, and Christopher R. Dickman, "Applying the Precautionary Principle to the Issue of Impacts by Pet Cats on Urban Wildlife," *Biological Conservation* 144 (2011): 1895–901.

2. Michael Woods, Robbie Mcdonald, and Stephen Harris, "Predation of Wildlife by Domestic Cats *Felis catus* in Great Britain," *Mammal Review* 33 (2003): 174–88; available online at tinyurl.com/ah6552e. This paper does not note that the survey was largely completed by children, nor does it provide any information about the format of the questionnaire used.

3. See, for example, Britta Tschanz, Daniel Hegglin, Sandra Gloor, and Fabio Bontadina, "Hunters and Non-hunters: Skewed Predation Rate by Domestic Cats in a Rural Village," *European Journal of Wildlife Research* 57 (2011): 597–602. The University of Georgia Kitty Cam project recorded 30 percent of outdoor cats capturing and killing prey; see www.wildlifeextra.com/go/news/domestic-cat-camera.html. This reduced to 15 percent if indoor-only cats were included.

4. See tinyurl.com/ak8c4ne.

5. Keen gardeners also seem to detest cats; a 2003 survey commissioned by the UK's Mammal Society as further grist for their campaign against cat ownership found that cats were rated alongside rats and moles as the mammals gardeners least like to see in their gardens.

6. Natalie Anglier, "That Cuddly Kitty Is Deadlier than You Think," *New York Times*, January 29, 2013, tinyurl.com/bb4nmpb; and Annalee Newitz, "Domestic Cats Are Destroying the Planet," *io9*, January 29, 2013, tinyurl.com/adhczar.

7. Ross Galbreath and Derek Brown, "The Tale of the Lighthouse-keeper's Cat: Discovery and Extinction of the Stephens Island Wren (*Traversia lyalli*)," *Notornis* 51 (2004): 193–200; available online at notornis.osnz.org.nz/system/files/Notornis_51_4_193.pdf.

8. B. J. Karl and H. A. Best, "Feral Cats on Stewart Island: Their Foods, and Their Effects on Kakapo," *New Zealand Journal of Zoology* 9 (1982): 287–93. Despite

this study, the cats on Stewart Island were subsequently exterminated, but (as predicted from the study) the kakapo continued to decline, and eventually scientists moved all the survivors to another, predator-free island.

9. Scott R. Loss, Tom Will, and Peter P. Marra, "The Impact of Free-ranging Domestic Cats on Wildlife of the United States," *Nature Communications* (2013): DOI: 10.1038/ncomms2380.

10. Cats may not be a recent introduction to Australia: it has been suggested that feral cats actually spread there several thousand years ago from Southeast Asia, following the same route as the dingo, the Australian feral dog. See Jonica Newby, *The Pact for Survival: Humans and Their Companion Animals* (Sydney, Australia: ABC Books, 1997), 193.

11. Christopher R. Dickman, "House Cats as Predators in the Australian Environment: Impacts and Management," *Human–Wildlife Conflicts* 3 (2009): 41–48.

12. Ibid.

13. Maggie Lilith, Michael Calver, and Mark Garkaklis, "Do Cat Restrictions Lead to Increased Species Diversity or Abundance of Small and Medium-sized Mammals in Remnant Urban Bushland?" *Pacific Conservation Biology* 16 (2010): 162–72.

14. James Fair, "The Hunter of Suburbia," *BBC Wildlife*, November 2010, 68–72; available online at www.discoverwildlife.com/british-wildlife/cats-and-wildlife-hunter-suburbia. The study Fair reports on was conducted by Rebecca Thomas at the University of Reading.

15. Loss, Will, and Marra, note 9.

16. Ibid.

17. Philip J. Baker, Susie E. Molony, Emma Stone, Innes C. Cuthill, and Stephen Harris, "Cats about Town: Is Predation by Free-ranging Pet Cats *Felis catus* Likely to Affect Urban Bird Populations?" *Ibis* 150, Suppl. 1 (2008): 86–99.

18. Andreas A. P. Møller and Juan D. Ibáñez-Álamo, "Escape Behaviour of Birds Provides Evidence of Predation Being Involved in Urbanization," *Animal Behaviour* 84 (2012): 341–48.

19. Eduardo A. Silva-Rodríguez and Kathryn E. Sieving, "Influence of Care of Domestic Carnivores on Their Predation on Vertebrates," *Conservation Biology* 25 (2011): 808–15. The cat and rat experiment was conducted in the early 1970s, when the ethics of animal experimentation were different than they are today; Robert E. Adamec, "The Interaction of Hunger and Preying in the Domestic Cat (*Felis catus*): An Adaptive Hierarchy?" *Behavioral Biology* 18 (1976): 263–72.

20. See the video at "Cat's Bibs Stop Them Killing Wildlife," Reuters, May 29, 2007; tinyurl.com/c9jfn36.

21. For more detailed advice, see www.rspb.org.uk/advice/gardening/unwanted visitors/cats/birdfriendly.aspx.

22. David Cameron Duffy and Paula Capece, "Biology and Impacts of Pacific Island Invasive Species. 7. The Domestic Cat (*Felis catus*)," *Pacific Science* 66 (2012): 173–212.

23. Andrew P. Beckerman, Michael Boots and Kevin J. Gaston, "Urban Bird Declines and the Fear of Cats," *Animal Conservation* 10 (2007): 320–325.

24. See www.rspb.org.uk/wildlife/birdguide/name/m/magpie/effect_on_songbirds .aspx.

25. James Childs, "Size-dependent Predation on Rats (*Rattus norvegicus*) by House Cats (*Felis catus*) in an Urban Setting," *Journal of Mammalogy* 67 (1986): 196–99.

Chapter 11

1. John W. S. Bradshaw, Rachel Casey, and Sarah Brown, *The Behaviour of the Domestic Cat*, 2nd ed. (Wallingford, UK: CAB International, 2012), chapter 11.

2. Darcy Spears, "Contact 13 Investigates: Teens Accused of Drowning Kitten Appear in Court," June 28, 2012, www.ktnv.com/news/local/160764205.html.

3. Summaries are available in several expert reports, including those commissioned by the Royal Society for the Prevention of Cruelty to Animals, the Associate Parliamentary Group for Animal Welfare, and the UK Kennel Club in partnership with the rehoming charity Dogs Trust. See www.rspca.org.uk/allaboutanimals/pets /dogs/health/pedigreedogs.

4. Domestic dogs have, of course, been selected for this trait throughout their association with man, since it is primarily their affection for us that makes them trainable.

5. Eileen Karsh, "Factors Influencing the Socialization of Cats to People," in *The Pet Connection: Its Influence on Our Health and Quality of Life*, ed. Robert K. Anderson, Benjamin L. Hart, and Lynette A. Hart (Minneapolis: University of Minnesota Press, 1984), 207–15.

6. You can find additional details of the introduction procedure on the Cats Protection website, www.cats.org.uk/cat-care/cat-care-faqs.

7. See the box in chapter 6 on clicker training. You can view a video of Dr. Sarah Ellis using clicker training to persuade a cat to walk into its cat carrier at www.fabcats.org/behaviour/training/videos.html (icatcare.org).

8. See Vicky Halls's article, for example, on the Feline Advisory Bureau website, www.fabcats.org/behaviour/scratching/article.html (icatcare.org).

9. For more information on declawing, written by a veterinarian, see "A Rational Look at Declawing from Jean Hofve, DVM" (2002), declaw.lisaviolet.com/declaw drjean2.html.

10. This constraint does not apply so much to dogs. Most pedigree dog breeds were originally derived from working types—terriers, herding dogs, guard-dogs, pack-hounds, and so on—and although the show-ring has diluted out much of their characteristic behavior, some still remains. Moreover, some working breed clubs have deliberately kept their lines separate to perpetuate the genes that enable their dogs to perform their traditional functions.

11. The Governing Council of the Cat Fancy, "The Story of the RagaMuffin Cat" (2012), www.gccfcats.org/breeds/ragamuffin.html.

12. Debbie Connolly, "Bengals as Pets" (2003), www.bengalcathelpline.co.uk /bengalsaspets.htm.

13. Charlotte Cameron-Beaumont, Sarah Lowe, and John Bradshaw, "Evidence Suggesting Preadaptation to Domestication throughout the Small Felidae," *Biological*

Journal of the Linnean Society 75 (2002): 361–66. This study included sixteen leopard cats and six Geoffroy's cats.

14. Susan Saulny, "What's Up, Pussycat? Whoa!" *New York Times*, May 12, 2005, www.nytimes.com/2005/05/12/fashion/thursdaystyles/12cats.html.

15. John W. S. Bradshaw, Giles F. Horsfield, John A. Allen, and Ian H. Robinson, "Feral Cats: Their Role in the Population Dynamics of *Felis catus*," *Applied Animal Behaviour Science* 65 (1999): 273–83.

16. In this context, neutering can be conceived of as a "meme," a concept that spreads, rather like a virus, from one human brain to another, producing biological consequences. See Susan J. Blackmore, *The Meme Machine* (New York: Oxford University Press, 1999).

Index